giving
natur
a hor

rspb

Kingfishers

David Chandler

B L O O M S B U R Y
LONDON · OXFORD · NEW YORK · NEW DELHI · SYDNEY

Bloomsbury Publishing

50 Bedford Square
London
WC1B 3DP
UK

1385 Broadway
New York
NY 10018
USA

www.bloomsbury.com

British Library Cataloguing-in-Publication Data
A catalogue record for this book is available from the British Library.

Library of Congress Cataloguing-in-Publication data has been applied for.

ISBN: PB: 978-1-4729-3367-6
ePDF: 978-1-4729-3372-0
ePub: 978-1-4729-3368-3

2 4 6 8 10 9 7 5 3 1

Design by Susan McIntyre
Printed in China by C&C Offset Printing Co.,Ltd.

To find out more about our authors and books visit www.bloomsbury.com. Here you will find extracts, author interviews, details of forthcoming events and the option to sign up for our newsletters.

giving
nature
a home

For all items sold, Bloomsbury Publishing will donate a minimum of 2% of the publisher's receipts from sales of licensed titles to RSPB Sales Ltd, the trading subsidiary of the RSPB. Subsequent sellers of this book are not commercial participators for the purpose of Part II of the Charities Act 1992.

Contents

Meet the Kingfisher

There is something about Kingfishers. No matter how many times you have seen them, they can still make a day special. The Kingfisher, or Common Kingfisher to be more precise, is one of our most colourful birds. It may not be easy to see one, but when you do you will probably know what it is, and will probably tell someone about the encounter, even if all you saw was little more than a blue blur. In Europe, Kingfisher identification is simple; with the exception of Turkey, Cyprus and Azerbaijan, there is only one kingfisher species. Its scientific name is *Alcedo atthis*. Alcedo means kingfisher, and 'our' Kingfisher is named after *Atthis*, a lady who lived on Lesbos around 600 BC.

It would be hard to confuse the Kingfisher with anything else and many people who have never seen one know what it looks like. They may not know how to tell a male from a female, or a juvenile from an adult, but even that is pretty straightforward given a half-decent view.

Opposite: An alert male. The distinctive Kingfisher is easy to recognise, but not so easy to see.

Below: The blue tips of its feathers work en masse to produce the Kingfisher's gloriously blue back and tail.

Small and colourful

The Kingfisher is definitely not an LBJ (or Little Brown Job, as many birders describe small drab birds they can't easily identify). It *is* little though. Measured from bill-tip to tail-tip, a Kingfisher is 16–17cm (6⅓–6⅔in) long, and its bill comprises roughly a quarter of that. It has short, rounded wings that span 24–26cm (9½–10¼in) when spread. An adult weighs between 34g and 46g (1¼–1⅔oz). To put this in perspective, at up to 100g (3½oz) a standard letter can weigh more than two large Kingfishers.

To describe this bird as blue and orange doesn't do justice to its stunning plumage, especially the range of blues, which in some lights can look green. Capturing these wonderful colours in words is not easy – it is hard to define the exact blue of a Kingfisher, not least because it isn't just one colour. The most striking blue is on their tail and back and can be seen particularly well on a bird in flight. It is only the tips of the feathers that produce this blast of colour, but because of the way the feathers

Below: Small, colourful and deadly – on this occasion at least.

overlap we just see the tips. This blue is described as electric and shifts from azure to cobalt.

The colour changes we observe are caused by the structure of the feathers rather than only by pigments. Their structure is such that when light hits them the blue wavelengths are scattered more than the red ones, a phenomenon known as the Tyndall effect. Where there

Above: It is no wonder that this beautiful bird is so popular. Its lovely plumage, seen here on a female, is most striking on the tail and back.

Left: Feathers from a juvenile Kingfisher's back (middle) and breast (left and right).

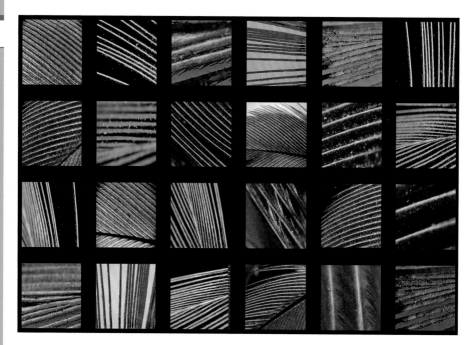

Above: Kingfisher back and tail feathers under the microscope. The shimmering electric blue is due to the structure of the feather, not just the pigmentation.

is electricity in the Kingfisher's plumage the blues are not static, and these areas can take on different tones in different light conditions, adding yet more sparkle. To be strictly accurate, the electric blue 'tail' is not actually the tail, but the upper tail-coverts. These are smaller feathers that cover the base of the tail, and in the Kingfisher's case, much of the rest of the tail too. The true tail lies underneath and is a darker blue.

Elsewhere on the bird there is less voltage. The blues on the wings and head are darker, and can appear green in some light conditions. The forehead, crown, nape, and the bar that descends from the base of the bill are brighter than the wings and head and some of the wing-coverts (small feathers that cover the bases of the flight feathers) are also tipped with a bolt of blue.

For one so bright the Kingfisher can be easy to overlook, and you may not notice it until your presence disturbs it and forces it to fly – low, fast and with whirring wings. Somehow those tropical colours don't always betray the bird's presence, and a perched bird can be surprisingly hard to spot among waterside vegetation.

Above and below: This perched Kingfisher looks conspicuous enough,
but they often go unnoticed until they take to the wing.

Male or female?

If you can see an adult bird well enough, male and female can be told apart, but not by their plumage. The clue is in the bill. If the bird is male, the bill is black. If it is female, the lower mandible is mostly reddish or orange, perhaps with a hint of brown, with the black restricted to the tip. The size of the black tip varies in females, but it is usual for it to cover one sixth to one quarter of the length of the bill, although it can be as much as two thirds, and on a few individuals the black is completely absent. Some males do have a bit of colour at the base of their lower mandible – ill-defined muted red or orangey blobs, or a suggestion of brown perhaps.

Right: With careful observation male and female Kingfishers can be told apart. The male (above) has a completely black bill, but the lower mandible of the female (below) has reddish or orange on it.

Adult or juvenile?

Juveniles look like muted adults, with more green, less blue, and blue-grey tones on the breast. When they fledge, both male and female juveniles have an off-white tip to an otherwise black bill. The pale tip remains for many months, and might still be visible the following spring, but not all of the black remains. Young birds leave the nest from the second half of May to the end of September, and about four weeks later, pinkish blobs appear at the base of the lower mandible. In the female, these get bigger and bigger and by January a chick's gender may be obvious, though it may take 8–12 months before the bill looks more or less adult. Working out the gender of a juvenile can be tricky while the lower mandible colour is developing, but the pinkish blobs on a male cover no more than a third of the bill.

A fresh-out-of-the-nest juvenile has black legs and feet. Those of the adults are almost fluorescent orange, and the slow, progressive brightening of a youngster's legs provides clues to its age. Normally the legs stay black for around 8–11 weeks. Then the red pigment surfaces, taking four to five months to transform black legs to glorious ruddy appendages.

Above: This is a juvenile Kingfisher. The plumage is duller than an adult's and there are darker tones on the breast. Its legs and feet aren't bright orange and there is a pale tip to the beak. This fledgling has been out of the nest for just three days.

The Kingfisher's new coat

Above: This young female is moulting for the first time. When the moult is complete she will have adult plumage.

Feather care is a vital aspect of any bird's life and every Kingfisher will devote a significant amount of time to keeping its plumage in good shape. But the feathers of even the most fastidious bird will need to be replaced from time to time.

The first moult takes juveniles to their first adult plumage. Kingfishers leave the nest during a period that can span more than four months of the year. As a consequence, some juveniles start moulting in July, but some have still not started by December. This is not a complete moult – it involves at least some of the feathers on the head and body and perhaps some tail feathers (more on some birds than others), but not the flight feathers. When winter closes in, moulting stops whether it has been completed or not. If necessary, the moult begins again in the spring. At this point some brand new inner flight feathers may appear too.

Adults moult after breeding, from June onwards. This involves most or all of their feathers, including the flight feathers, though the body feathers are the first to be renewed. Primary feathers are not shed until the new ones are well grown. By the end of November, if not before, the moult may be completed, though as in the post-juvenile moult, it may grind to a halt as winter approaches and be continued in the next year. It is not unusual for the moult of one wing to be out of synch with the other. The tail moult may be 'unbalanced' too, and it seems that the older the bird gets, the more asymmetric its moult becomes. Unless it is really cold, body feathers may be lost and renewed at almost any time, and some plumage may be moulted twice a year.

Above and right: Kingfishers may devote up to two hours a day to preening their feathers, using their bill to apply oil from the preen gland, which is on the bird's rump. Diligent care of feathers is vital for birds that are frequently immersed in water when they dive for food.

How common is the Common Kingfisher?

Above: The Common Kingfisher has an extensive range. This one was photographed in Nepal.

BirdLife International estimates the world population of Common Kingfishers at fewer than 600,000 birds and the European population at 97,500 to 167,000 pairs. The species lives in lots of different countries and, globally speaking, conservation scientists see little cause for concern.

At a European level, however, things are not quite as rosy. Because of a rapid drop in numbers the Kingfisher is classied as 'Vulnerable' on the European Red List.

Closer to home, *Birds of Conservation Concern 4*, published by a consortium of conservation organisations in 2015, summarises the population status of the UK's bird species and that report places the Kingfisher on its Amber list, signifying that the species is of moderate conservation concern. Red-listed birds are those most urgently in need of help.

The UK breeding population is estimated at 3,800–4,600 pairs. For comparison, the UK is also home to about 200 pairs of Ospreys (*Pandion haliaetus*), and about 7.7 million Eurasian Wren (*Troglodytes troglodytes*) territories. (The Wren's population is expressed as territories because one male Wren may breed with more than one female in his territory.)

Different looks and different places

The Kingfisher is not just a bird of Europe. Its range extends from Ireland, Portugal and Morocco in the west, eastwards through continental Europe and discontinuously across Asia as far as Japan and Sakhalin. Birds in southern Sweden, Finland and adjacent parts of Russia are the northernmost breeders. Further south, you can find Common Kingfishers in India and China and down into Thailand and Malaysia. Keep going south and you can still see the same kingfisher species as you see in Europe – they are found in Indonesia, the Philippines and Papua New Guinea. You can even find them in the Solomon Islands. But you won't find any in the Americas, Australia, or sub-Saharan Africa – you will have to content yourself with other kingfisher species in these areas.

Below: The two European sub-species of Kingfisher. They are not easy to tell apart!

The Common Kingfisher comes in more than one form. The International Ornithological Congress (IOC) list of world bird names includes seven different subspecies, each denoted by a different suffix to the two-part specific name. In Europe there are two subspecies, *Alcedo atthis atthis* (the 'nominate' form), and *Alcedo attis ispida*. *Alcedo atthis atthis* is found around the Mediterranean, in south-east Europe and in European Russia. Kingfishers seen further north and west are of the subspecies *ispida*. The average birdwatcher, or even the above average birdwatcher, might wonder what all the fuss is about. There are minor differences in colour and size, but the difference can be as small as half a millimetre of wing-length, and even the best binoculars in the world will not spot that. To make it even harder, that wing-length difference is not totally reliable.

The Kingfisher Family

There may not be many kingfisher species in Europe, but take a global view and it's a very different story. This chapter provides an overview of the kingfisher family as a whole, and, briefly, of some of their close relatives. In most of this book I have used 'Kingfisher' instead of 'Common Kingfisher' (its more correct name) because that is what most people say in everyday speech. In this chapter, however, 'kingfisher' is also used to encompass other kingfisher species.

Its more formal common name and its scientific name distinguish the Common Kingfisher from 112 other kingfisher species that can be found around the world. Exactly how many other kingfisher species there are is uncertain – bird taxonomy doesn't stay still for long and different authorities have reached different conclusions.

It is not just the number of species that differs. The International Ornithological Congress (IOC) puts the kingfishers in one family, the Alcedinidae, alongside two other families in the order Coraciiformes: the rollers (Coraciidae), and the ground-rollers (Brachypteraciidae) of Madagascar. The kingfishers total many more species than both of the other families put together (17 species).

Opposite: 'Our' Kingfisher after a failed mission – there is no fish in the beak.

Below left: A Long-tailed Ground-roller; one of five ground-roller species, all of which are found only in Madagascar. None of the others have a long tail.

Below: A European Roller. Rollers are so-named because of their rolling courtship flights.

An alternative classification has treated the Alcedinidae as not one family, but three: the river kingfishers (Alcedinidae), the tree kingfishers (Halcyonidae) and the water kingfishers (Cerylidae). The species' names and taxonomy that I have used in this chapter follow the IOC classification.

Right: The Sacred Kingfisher is found in Australasia, Fiji and Indonesia. It is not as spectacularly coloured as some other members of the family but has the characteristic body shape and a powerful bill. The diet of this species includes a variety of small vertebrates and arthropods, including crabs.

Structure and colour

The kingfisher 'design template' includes a big head on a chunky, short-necked body, short legs, a stumpy tail and a long, dagger-like bill. They have very good vision whether above or below the water surface, thanks, in part at least, to elliptical lenses. At only 10cm (4in) from bill-tip to tail-tip, and tipping the scales at a mere 10g (⅓oz) or so, the aptly named African Dwarf Kingfisher (*Ispidina lecontei*) really is the dwarf of the family, not to be confused with the similarly named, but larger, African Pygmy Kingfisher (*Ispidina picta*), which is 2cm (¾in) longer. The Giant Kingfisher (*Megaceryle maxima*) is at the other end of the scale. Found in much of sub-Saharan Africa, at 45cm (18in) long and with a mean weight of 355g (13oz), this is Africa's heavyweight. But the real giants among kingfishers are in Australasia. The Laughing Kookaburra (*Dacelo novaeguineae*) is a similar length to the Giant Kingfisher, but can reach 500g (18oz).

The kingfisher's bill is one of its most striking features. As a vital feeding tool, its structure indicates something of the owner's diet. A bird that eats mostly fish, for example Australasia's Little Kingfisher (*Ceyx pusillus*), has a pointed bill that looks like it has been squeezed at the sides, while one that prefers to dine on insects and other invertebrates, such as Africa's Striped Kingfisher (*Halcyon chelicuti*), which feeds mainly on grasshoppers, has a shorter, flatter bill. No other kingfisher has a bill like

Top: Australasia's Little Kingfisher is primarily a fish-eater. It is only found in parts of the Northern Territory and Queensland.

Above: The Brown-hooded Kingfisher is not a fish-eater. This African species has a flatter bill than the fish-eating kingfishers.

Above: A kingfisher's bill is not just a feeding tool. Some species use their bill to excavate a nest tunnel, as this bird appears to have been doing.

Below: Africa's Blue-breasted Kingfisher (*Halcyon malimbica*), a species found in mangroves and forested areas.

that of the Shovel-billed Kookaburra (*Clytoceyx rex*). This unusual bird, found only in New Guinea, sits in a genus all of its own. It is big, measuring 33cm (13in) in length, and sports a stout, squat, cone-shaped bill that is clearly not the bill of a fish-eater. In fact, it forages in damp soil and mud for earthworms, insects, snails and lizards.

At the other end of the body, kingfishers have weak feet with three or four toes. A four-toed kingfisher has three toes pointing forwards and one pointing backwards, an arrangement common to most birds. Three-toed species lack the innermost forward-facing toe. All kingfishers have feet that are syndactyl; on each foot, two of the toes are joined for some of their length. The middle and outer toes are joined for more than half their length, and, if an inner toe is present, about a third of that is fused to the middle toe too.

The family has been lavishly adorned from the colour palette, and many species are a combination of blue and orange. References to blue are a recurrent theme in their common names: Lazuli Kingfisher (*Todiramphus lazuli*), Ultramarine Kingfisher (*Todiramphus leucopygius*), Azure Kingfisher (*Ceyx azureus*), Malachite Kingfisher (*Corythornis cristatus*) and Cerulean Kingfisher (*Alcedo coerulescens*), and numerous mentions of plain old blue,

including blue-breasted, blue-banded, shining-blue and blue-eared. The blue-and-orange theme can be seen on kingfishers in many parts of the world, including Asia, Africa and Australia, but those in the *Ceryle, Chloroceryle* and *Megaceryle* genera are painted from a different palette. North America's Belted Kingfisher (*Megaceryle alcyon*) and the Pied Kingfisher (*Ceryle rudis*) of Africa

Above: Kingfisher blues – Azure Kingfisher (left), an Australasian species, and Cerulean Kingfisher (right), a species from Sumatra and various islands off Australia's northern coast.

Below: Asia's Blue-eared Kingfisher (*Alcedo meninting*) Note the lack of orange behind the eye compared to 'our' Kingfisher.

Above: Malachite Kingfisher, a small, common and widespread African species.

and Asia are good examples. The males and females of most kingfisher species look similar, and juveniles are a toned down version of the adults. Paradise kingfishers (*Tanysiptera* species) are an exception – the drab juveniles look quite different to the adult birds.

Right: This is a White-throated Kingfisher, a mainly Asian species whose range just stretches into Europe.

Above: A Belted Kingfisher from the Americas.

Little did you know ...

The kingfisher family as a whole may not be quite as you imagine it …

- Not all kingfisher species live by rivers and lakes. Some live a long way from water, and more than half of them live in forested areas.

- Not all members of the family have 'kingfisher' in their name, though the vast majority do. The exceptions are the five species of Australasian kookaburras.

- For some species, 'fisher' is a misnomer. None of them survive on fish alone, and some species, such as the Shovel-billed Kookaburra (*Clytoceyx rex*), get most of their sustenance from terrestrial invertebrates.

- Not all kingfishers are diurnal (active during daylight hours). New Guinea's Hook-billed Kingfisher (*Melidora macrorrhina*) is active at dawn and dusk and perhaps at night.

- Bird ringers have discovered that when a Common Kingfisher is being held in the hand, if it is laid on its back it will make no attempt to fly off.

Laughing Kookaburra.

Blue-winged Kookaburra.

Where do they live?

Above: It's obvious why this Asian species is called a Stork-billed Kingfisher (*Pelargopsis capensis*).

Below: This is a Chattering Kingfisher – an appropriate name for a bird that lives on the Society Islands!

Kingfishers are found on every continent apart from Antarctica, and are also absent from far northern latitudes. To see the greatest diversity of kingfishers, head to Australasia or the tropics of Africa and Asia. Cameroon is home to 12 species from six different genera. Borneo hosts 11 species in seven genera and Australia has 11 species in five different genera. There are even kingfishers on some very remote Pacific islands, including the Mewing Kingfisher (*Todiramphus ruficollaris*) of the Cook Islands and the Chattering Kingfisher (*T. tutus*) on the Society Islands. In terms of kingfisher diversity, Europe is definitely impoverished. If you include the two records of Belted Kingfisher between 1950 and 2007, the British kingfisher lists peaks at just two species from two genera, and it is only the Common Kingfisher that occurs regularly. Two more species can be added for Europe as a whole, with small populations of White-throated Kingfisher (*Halcyon smyrnensis*) resident in Turkey and Azerbaijan, and populations of Pied Kingfisher (*Ceryle rudis*) living in Turkey and Cyprus.

Kingfishers are not found in every habitat in the countries in which they occur, but they do occur in a wide range of habitats. Many people think of kingfishers as birds of lakes and rivers, and while this is accurate for some species, for many it is not. Kingfishers live on coral atolls, up mountains, in mangroves, deep in rainforests, on seashores and in savannas. Perhaps surprisingly, more species live in forests, with or without streams, than in other habitats. Less surprisingly, most deserts are kingfisher-free zones, but even this is not a universal truth – in Australia, the Red-backed Kingfisher (*Todiramphus pyrrhopygius*) survives in seriously arid desert.

Above: The Red-backed Kingfisher is found in a variety of habitats across most of Australia, including desert areas.

Above: An African Dwarf Kingfisher weighs about 10g (⅓oz) and is the world's smallest kingfisher species.

What do they eat?

Contrary to what you might think, most species don't depend on fish. Some do take fish, but the family's diet sheet lists a host of invertebrates including crabs, earthworms, molluscs, centipedes, spiders, various insects and an impressive range of non-fish vertebrates too, including small mammals, tadpoles, frogs, birds, snakes and other reptiles. The Belted Kingfisher (*Megaceryle alcyon*) even feeds on plant material occasionally. Generally speaking, waterside kingfishers are primarily fish-eaters, and species that live in dry places eat other things. Some species are picky eaters,

Above: The Green Kingfisher is found in South and Central America and some parts of the southern United States. It lives up to the 'fisher' part of its name – some of its kingfisher relatives have much less fishy diets.

Opposite: A Common Kingfisher in action underwater.

Below: The Common Kingfisher doesn't feast only on fish. This bird has caught an insect larva.

with a limited range of prey items making up most of their diet. The Pied Kingfisher (*Ceryle rudis*), which is primarily a bird of Africa, Asia and the Middle East, and the Green Kingfisher (*Chloroceryle americana*) of the Americas are examples of species that eat mainly fish. Others are much less discerning. The Sacred Kingfisher (*Todiramphus sanctus*), for example, which is found in Australasia and Indonesia and on various Pacific Islands, will eat a host of invertebrates, as well as reptiles, amphibians, birds, mammals and fish. Even the specialist fishers, such as the Common Kingfisher, supplement their fishy meals with invertebrates.

The Common Kingfisher's familiar perch–watch–dive–perch approach to hunting is not the only technique

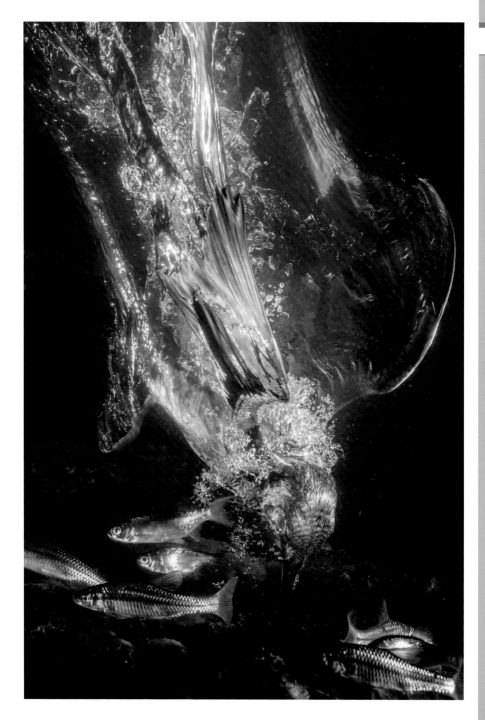

Above: Only a handful of kingfisher species can hover. This is a Pied Kingfisher.

the kingfisher family uses to find the next meal. The family's food-gathering skills include diving from a hover, which gives a Pied Kingfisher access to prey up to 2m (6½ft) beneath the water's surface. This species, and the Belted Kingfisher, hover up to 12m (39ft) above the surface of the water. Hovering is not widespread among kingfishers, however – only seven species can do it and these are the most fish-dependent species, including the Common Kingfisher. Despite the family name, many species, including those in the *Tanysiptera* and *Halcyon* genera, and the *Dacelo* kookaburras, feed terrestrially, though they may take food from the water surface too. Their technique is a bit like the aquatic fishers, but with a swoop rather than a dive, and prey is picked off the ground or plants, including trees. Some species – Africa's *Halcyon* kingfishers, for example – snatch insects in flight. Australia's Red-backed Kingfishers have even been caught in the act of breaking into the mud nests of Fairy Martins (*Petrochelidon ariel*) to steal some young avian protein. Laughing Kookaburras can be thieves too, using surprise as a tactic to grab food from the hands of picnicking humans. They will take snakes too.

Some food items receive special treatment before swallowing. Among fish-eaters, the *coup de grâce* may be administered by bashing the victim against a perch, and this can break bones and make sharp spines safer to swallow. Similar behaviour is also exhibited by other kingfishers, with terrestrial prey being bashed against a perch or on the ground.

The mating game

Above: *Halcyon* kingfishers will sing to attract a mate. This is a Brown-hooded Kingfisher.

Typically, kingfishers are monogamous. Most can probably breed when they are one year old and will seek to evict others of their species from their territory, sometimes with considerable determination. Elaborate courtship displays do not seem to be common in kingfishers, though those in the *Halcyon* genus will belt out a song from the top of a tree, twisting from side to side with wings held open.

Kingfishers nest in holes – in trees, termite nests and earth banks. Tree-nesters include Australia's Laughing Kookaburra and Africa's Woodland Kingfisher (*Halcyon senegalensis*), which will move into a woodpecker or barbet hole. Many forest-dwelling kingfishers use termite nests (termitaria) as their nest sites, digging into the insects' handiwork at ground level. The earth bank-nesters, including the Common Kingfisher, excavate a tunnel in waterside banks, including man-made banks and may take advantage of waterside holes in concrete too. With a tunnel of 8.5m long (over 27ft) the Giant Kingfisher holds the record for the longest nest tunnel of any kingfisher. Creating a nest site can be dangerous for a kingfisher. Sometimes, as work commences, a bird will launch itself at what it hopes will become the nest entrance with such zeal that he or she is mortally wounded.

Below: The Giant Kingfisher is Africa's heaviest kingfisher, and a record-breaking tunnel excavator!

Whatever the nest site, kingfishers lay shiny, white eggs that are incubated by both parents, though scientists think that, generally, the female does the bulk of the work. Within the family, clutch sizes range from one to ten eggs. Species of temperate latitudes have bigger clutches, averaging roughly 6–7 eggs, than those of tropical regions, which average 3–4 eggs. This is a general phenomenon in the bird world, and may be because fewer birds survive the temperate winter than the tropical non-breeding season, which, coupled with a spring abundance of food, means that each pair can feed a larger brood. Typically, the eggs don't all hatch at the same time – those that were laid first hatch first. This means that when the male and female bring food to their young, they will be feeding chicks of different sizes and different ages. The clutch of 4–8 Common Kingfisher eggs, however, hatches more or less synchronously.

The challenges of raising young are made easier for Pied Kingfishers and Laughing Kookaburras by assistance from 'helpers' – individuals of the same species that help the parent birds (see also box, below).

Pied Kingfisher (*Ceryle rudis*)

Pied Kingfishers are found in Africa and Asia, including the Middle East, with smaller numbers in south-east Europe. This species is unique among kingfishers in being a colonial nester. Normally this occurs when suitable nesting habitat is hard to come by and is not alongside the feeding area. Some are solitary breeders, however, and will stake their claim to a length of river where they build their nest and fish.

Parents benefit from the contributions of primary and/ or secondary 'helpers'. Typically, a primary helper is a one-year-old male offspring of one or more of the parents. Secondary helpers are thought to be unrelated birds that are either not breeding or that tried to breed earlier that year but failed.

Helpers assist with feeding the breeding birds and their young, and provide extra muscle to drive off interlopers or predators. A pair may benefit from the contributions of several secondary helpers but normally has just one primary helper.

Home and Away: Habitat and Movements

When it comes to habitat selection, the Kingfisher is not as fussy as you might think. The key requirements are water, plenty of fish, and, ideally, some perches to catch them from. These criteria apply all year round, with an additional breeding season requirement of somewhere to nest.

Kingfishers prefer still water or water with little flow. They avoid murky water, and usually opt for freshwater habitats, particularly in the breeding season. They are very capable at catching fish but their small size limits the size of fish they can tackle. A 10cm (4in) fish is about as much as the Kingfisher can manage. A good habitat has lots of fish that are smaller than this.

Opposite: Kingfisher habitat can be very close to people. This female Kingfisher could live at the river at the end of the garden.

Below: If you are lucky enough to live by a river, you may find Kingfishers right on your doorstep. This male has found a convenient perch at the edge of a riverside garden.

Right: This river is ideal Kingfisher habitat: clear water, plenty of perches and soft banks that are easy to tunnel into.

Right and below: Kingfishers need water. Rivers, canals (right) and lakes (below) can all provide good Kingfisher habitat.

An ideal breeding habitat will have a nest site near the water, typically a more or less vertical, or even overhanging, bank that is soft enough for tunnelling. It is not the norm, but if there is no alternative, nest sites can be more than 250m (820ft) away from where the birds feed.

Kingfishers can be found around rivers, canals, ditches, streams and lakes. They like some cover at the water's edge and perches to fish from, and prefer shallow water to deep water, presumably because it's easier to fish in shallow water. Despite their elusive character, they can sometimes be found very near human habitation. They will take shortcuts too, and may be seen flying across fields to save going all the way around a river bend.

This is not a bird of crashing, bubbling upland waterways, and does best in clean, unpolluted habitats. It is primarily a bird of the lowlands. Few live more than 650m (2,100ft) above sea level and finding one anywhere in Europe higher than 900m (3,000ft) above sea level is very rare, though some have been found in Russia at around 2,000m (6,500ft) above sea level.

Above: Some Kingfishers spend the winter at the coast, and may even fish in the sea.

Wintering habitats

Some Kingfishers move to different areas for the winter. In the United Kingdom, most birds probably do not, though there are birds that migrate to the coast for winter and a few have crossed the sea to mainland Europe. Outside the breeding season, the species is more cosmopolitan in its choice of habitat and can be found in a wider range of watery environments. When not constrained by the need for a suitable nest site, some will forage in brackish water, and sometimes even in the sea. David Boag (author of *The Kingfisher*, first published in 1982) suggested that it was mainly immature birds taking fish from saline waters, and that these were probably birds that did not have a winter feeding territory.

Do they migrate?

Some Kingfishers migrate, and others stay put. When they do migrate, most do it under cover of darkness. Ornithologists believe that adults are less likely to seek

Below and opposite: Some northern European Kingfishers migrate south and west to avoid cold winter weather, but not all of them do. A harsh winter can be serious if fishing patches freeze over.

sunnier climes than juveniles, and if they do move they will not travel as far. Adult males are more sedentary than adult females.

Clearly when your feeding area turns to ice, it makes sense to move on. Populations that breed in Poland, the Baltic States and Finland, and further east, across Russia as far as China and Sakhalin, are migratory. They spend the winter in more agreeable climates in Europe, the Middle East and North Africa, with some from the far east of the global range travelling as far as the Philippines and Indonesia.

Most European migrants head south and west to spend the winter in less demanding parts of the breeding range. Some go further and do their winter fishing in the Mediterranean or in North Africa, including areas where Kingfishers do not normally breed. Others make it to the Middle East and, even more remarkably, to India and Pakistan.

Birds in central Europe are 'partial migrants': some migrate and some do not. Those that do head for gentler weather to the south and west, leaving those that stay to take the risk of falling victim to an unusually harsh winter.

Above: If their normal fishing patch evaporates, some Mediterranean Kingfishers fly to the coast.

Kingfishers that breed further south can be pretty sedentary, though some around the Mediterranean head for the coast when summer heat dries out their favourite fishing patches. Those that breed in the UK do not migrate, although the few birds that manage to find somewhere to breed in the uplands may abandon their territories in winter. Data from bird ringing suggest that these birds travel no more than 25km (15½ miles) from their breeding territory, and that individuals in the north of the country move a bit further than those in the south.

Germans call the Common Kingfisher the Eisvogel ('ice bird'), because it moves south when waters freeze over. The Dutch name, Ijsvogel, has the same meaning. In the UK, severe winter weather can turn favourite fishing patches to ice and birds may have to go elsewhere to find food, frequently moving towards the coast. There is not much reliable data on this winter movement – ringing has thrown very little light on the phenomenon. When the birds do move, they run the risk of entering another Kingfisher's territory, where they will not be welcome.

How far do they travel?

Ringing birds gives us some clues about how far kingfishers travel, though only a small proportion of the birds that are ringed are found again – in the British ringing scheme, over three quarters of the birds that were found again were dead; the others were recaught. A 1964 publication tells of two adults ringed in the former Czechoslovakia that took a southerly course to Italy, covering 790km (490 miles). German birds have been recovered in Belgium, Denmark, France, Spain and Italy. Some continental birds find their way across the sea from time to time, and pass through, or spend the winter in, the UK. Birds from France, the Netherlands and Germany have all been recorded in Britain. A youngster from Belgium was recovered in the UK and a bird showed up in Wales eight months after it was ringed in Brittany. Even more remarkable is a bird ringed in Wales that was recaught in northern Spain less than three weeks later. Kingfishers ringed in Britain have also been found in France and Belgium.

Below: Frozen waters make life difficult for Kingfishers.

Right: Ringing has yielded some fascinating information about Kingfisher movements.

The long-haul experts seem to be the juveniles. Adults do not tolerate their young for very long, and they are soon ousted from the breeding territory. A study in Russia showed youngsters moving up to 30km (19 miles) a day, and a young upstart from the former Czechoslovakia completed a marathon journey of 1,820km (1,340 miles) to south-east Spain. Other birds from the same part of Europe have made it to Malta and Sicily. The Russian study also found Kingfishers holding their first breeding territories no more than 20km (12½ miles) from where they hatched. In other areas distances of up to 128km (80 miles) are known. In Britain and Ireland, ringing data show that birds moved an average of just 9km (5¾ miles) in the year after hatching and the adults seem even less adventurous – recoveries of dead adults average a mere 3km (1¾ miles) from the place they were ringed.

In surprising places

Kingfishers have been recorded in unexpected places. In March 1939, one showed up on the Skerries, a group of islands north-west of Anglesey. From time to time they are found on Scotland's Western or Northern Isles. These are places where they do not normally occur and records there may be of migrant birds from mainland Europe.

Sometimes the Kingfisher is seen in woodland, an unusual habitat for this species. My most unlikely sighting happened one June over 25 years ago when I was driving up a slip road onto the M11. A blue blur was flying to my left, more or less parallel to the road.

Above: It may be near water, but this is not where you would expect to see a Kingfisher.

Below: There's nothing particularly surprising about the location of this Kingfisher.

Catching Fish ... and More

The Kingfisher lives up to its name. Mostly, it eats fish, and the list of species it consumes is impressive, including Bullhead, Minnow, Three-spined Stickleback, loaches, Grayling, Carp, Perch, Pike and plenty more. As noted earlier, there is a size limit – a large Pike does not need to fear for its life when a kingfisher flies over. A small one might need to be careful though.

Kingfishers like their fish alive, lean and not too long – up to 7/8 centimetres (2¾in) long is good, and even fish 10cm (4in) in length may be taken. David Boag attempted to work out how many fish a Kingfisher eats. To do this he put Minnows into an artificial pool, which they couldn't escape from, in a natural waterway. When he was reasonably confident that a bird was getting its sustenance almost entirely from that pool he could work out how many fish the bird was taking because he knew how many he was putting in. After 11 days of feeding, he

Opposite: Success!

Below: Minnows and Sticklebacks make ideal food for Kingfishers.

Above: It may look like the result of a successful fishing trip, but in manipulating its catch, this bird is likely to lose all but its innermost victim.

concluded that a bird busy getting to know a potential mate had an average daily intake of 17.8 fish. This equates to about 46g (1²/₃oz) of Minnow-meat, just a little more than the weight of the bird. The Kingfisher ate more fish when the sun shone, and less when it was cloudy.

Another study collected data on the number of different types of fish eaten by Kingfishers in Belgium. About 36 per cent of the diet was fish in the carp family. Three-spined Sticklebacks accounted for nearly 29 per cent, with Bullheads being taken almost as often (about 28 per cent of the diet). Other prey included Perch and trout. Further east, birds in Estonia are noted as particularly fond of Gudgeon, Bleak and Minnows. In the UK, work by wildlife film-makers Ron and Rosemary Eastman concluded that the birds they studied around the River Test in Hampshire devoured mostly Minnows, Sticklebacks and Bullheads.

The Eastmans and David Boag did separate experiments to try and tempt Kingfishers to take non-fish prey items by providing alternatives where they could be very easily caught – the Eastmans put theirs in a tin bath in a stream. Their experiments were limited, with earthworms, Smooth Newts, crayfish and dragonfly larvae all declined. When an elver Eel was offered the bird had a go, but was unsuccessful and dropped it.

The seafood diet

Information on the gastronomic delights enjoyed by
Kingfishers at the coast is harder to come by and earlier
works have presumed that molluscs and crustaceans make
up the bulk of the diet. Coastal Kingfishers do take saltwater
fish including goby, Tompot Blenny, Cornish Sucker and
Sand Smelt, and they are also known to eat prawns.

Not fish again

Despite a predilection for all things fishy, Kingfishers are
sometimes more adventurous in their eating habits. From
time to time, frogs are taken, much to the surprise of a
certain Mr Topp, a taxidermist in Berkshire, who, according
to a note in Volume 10 of *British Birds* (a monthly journal
for keen bird watchers), found a rather large one inside
a dead Kingfisher – the frog's back leg was 6.4cm (2½in)
long! Kingfishers that live in the more arid parts of the
Algarve take reptiles. Invertebrates aren't always safe either,
and that includes the occasional land-based insect. Bugs,
beetles, spiders, stoneflies, mayflies, shrimps and the larvae
and adults of caddis flies and dragonflies have all ended
up inside the Kingfisher. Work published in 2015 (by Čech
and Čech), however, concluded that any non-fish items in
a Kingfisher's diet, which in their study included dragonfly
larvae and Great Diving Beetle larvae, were probably

Left: Kingfishers eat frogs
occasionally too.

Above and right: Kingfishers sometimes eat Great Diving Beetle larvae (above) and dragonfly larvae (right) but probably by mistake.

caught by mistake. Their Kingfisher menu included the occasional crayfish (in this case Spiny-cheek Crayfish), newt and a lizard.

There are even records of Kingfishers going for bits of submerged plants, and during the very cold winter of 1859–60 one opportunistic individual in Norfolk fed on suet that had been put on a bird table. There is also a record from Devon of an individual that took suet, and was then fed from the bird table with bits of fish.

Skuas are infamous for their piratical stealing of food from other birds. Occasionally, Kingfishers play this game too and have been seen depriving the White-throated Dipper *(Cinclus cinclus)* of its hard-won catch. This despicable behaviour is not restricted to birds either – Water Shrews *(Neomys fodiens)* have also had their prey stolen.

Below and right: Skuas (below) are well known for it, but Kingfishers take food from other birds too. The White-throated Dipper (right) is a known victim.

Cough it up

It is not just owls that regurgitate pellets (undigested masses of food remains, such as bones, fur and some insect parts). Other birds do too, including the Kingfisher. To find some, first find a Kingfisher's favourite perch – look for conspicuous white streaking from the bird's excrement. The area underneath the perch is a good place to look for pellets. Lots of the pellets land in the water but some land at the waterside. They disintegrate quickly, but a reasonably fresh pellet might still be more or less intact. Cameraman and photographer Ian Llewellyn, who took many of the pictures in this book, was able to watch the coughing up process. The bird gave up any attempts at fishing and sat fairly still for about 45 minutes before finally ejecting the pellet, which it shook out of the side of its beak. Once the pellet was out, the bird went fishing again. Ian also managed to retrieve a pellet very soon after it was regurgitated – a bird landed outside his hide, coughed up a pellet and flew off soon afterwards. The pellet was held together by a smooth coating and was covered in tiny bubbles, presumably giving the bird some protection from sharp bones.

Ian also found a pellet that he believed was produced by a bird that had struggled to find fish during a cold spell in January 2009. He had seen that the bird's hunting now included quick, short dives very near the river bank and suspected that insects and very small Sticklebacks were being taken. Sure enough, when the pellet was examined, Ian found insect parts, which he believed were from a water boatman – these are illustrated on the right.

Above and below: Pellets may be found close to Kingfisher perches. They usually consist of fish bones as seen in the two pictures above. The picture below shows insect parts from a pellet ejected during a cold winter spell.

A particularly dedicated ornithologist, or perhaps it was more than one, searched through over 14,000 bits of fish remains that had been deposited in pellets by some Belgian Kingfishers. Most of the remains (90 per cent) were from some sort of carp, or a Bullhead. This work was described in a paper by Hallet, published in 1977.

Catching ...

The Kingfisher is a dive-fisher *par excellence*. Experienced birds can enjoy a high success rate – roughly four out of five dives are successful, and with this hit rate, the bird probably doesn't have to spend a large part of the day feeding. Work in Nepal, however, came up with a hit rate of just below 40 per cent. Kingfishers often dive from a perch, and they prefer one that doesn't move much, and ideally, not at all. Normally, a fishing perch is 1 to 3m (3¼–9¾ft) above the water's surface, but some are higher. A note to *British Birds* from R. A. Frost tells of a particularly intrepid bird that successfully fished the River Trent in England from an electricity cable roughly 11m (36ft) above the water. Ian Llewellyn has seen a successful dive into mini-rapids, with the bird popping up about 30cm (1ft) downstream.

The dive itself is remarkably speedy. Boag measured a bird's 1m (3¼ft) dive as taking about 1.2 seconds from perch to water to perch (with fish). When a Kingfisher needs a bit more momentum than a straightforward perch-dive provides, if targeting a deeper fish for

Below: Despite its dart-like appearance, the Kingfisher does not spear its prey. Occasionally fish become impaled on the bill, but if this happens the meal will usually go to waste because the Kingfisher cannot reposition the fish for swallowing.

example, it will fly up to begin the dive from a greater height. Powering wings add speed to the dive, the tail makes any final tweaks to its course, and the wings are pulled up over the bird just before impact. Mostly, the species is not a deep-diver, taking fish from just below the surface, to a depth of about 25cm (10in). Diving to about 1m (3¼ft) beneath the surface is not unknown, however, and deeper fish need more vertical dives. Particularly shallow dives require a different technique. The head is retracted so that the bill barely protrudes beyond the chest, and partly spread wings help to reduce the depth of the dive. Sometimes, the Kingfisher forces hidden fish into open water by 'belly-flopping'. Dinner is served shortly afterwards.

When a diving Kingfisher enters its watery hunting ground, its eyes are closed but the bill is agape. Its underwater course has already been determined and no more adjustments are made. The bird's instinctive three-dimensional calculations of fish position, presumably informed by past successes and failures, are remarkable, and the Kingfisher can be unerringly accurate. The task now is to get back to the surface, a job made easier by beating wings and feather-trapped air, but with a

Below: The shallow angle of entry of this diving Kingfisher suggests that it wasn't targeting a particularly deep fish.

final hurdle of breaking out of the water completely. Normally, an experienced bird leaves the water behind without problem, but sometimes a bird will struggle, only liberating itself after several considerable efforts.

This dive-fisher does not need a perch to dive from. In the absence of a solid launch pad, the Kingfisher hovers. It is not true hovering – if there is no breeze at all the bird cannot perform this trick. Hover-fishing is often employed around the coast and on estuaries. Hovering also enables the Kingfisher to snatch perched dragonflies from reedbeds (this has been observed in Spain) and they may even take sizeable spiders when they come out of their sand dune burrows. Hovering birds have also been seen doing Spotted Flycatcher (*Muscicapa striata*) impressions – with airborne insects as would-be prey – and a perched juvenile has been seen mimicking the insect-hawking attempts of other birds. Records of Kingfishers interrupting 'normal' flight with an 'in-flight' dive win the blue blur even more respect as a king among fishers.

Opposite: The Kingfisher pulls its wings up just before it hits the water.

Above: Getting out of the water is the final challenge of a Kingfisher dive.

Below: When there is no solid perch from which to launch, the Kingfisher can dive from a hover. Unlike a 'true' hover, this kind of flight cannot be achieved without the help of a breeze.

... and despatching

Different prey items require different 'preparation' before they are swallowed. In essence, the despatch technique involves repositioning the fish so that it is gripped in the bird's bill a little forward of the tail, and quickly swinging it to the side to whack its head on something solid, typically a branch.

Sometimes the despatch is simple. A Minnow might need just one or two whacks before the squirming fish goes down the throat, or even none at all. Fish with spines or spiny fins, Sticklebacks for example, need more care for obvious reasons. Their demise must be well and truly complete before they are swallowed, or the spines still stick out. Inexperienced birds are sometimes too hasty, swallow too soon, and cough the uncomfortable meal up again before giving it a more thorough beating.

The Bullhead has substantial pectoral fins and it too requires extensive preparation before it is safe to eat. This fish is a choking hazard if the fins are sticking out. Beating does the trick. David Boag saw one determined bird whack a Bullhead 30 times. This wasn't enough. The fish went into the bird and back out again. Not once, but ten times. It took 57 more whacks to complete the job. The total preparation time was 13 minutes.

A vigorous head-shake by the successful hunter, shortly after the fish is swallowed, is the norm.

Below and opposite: Kingfishers hit their unfortunate prey against something solid, often a branch, before carefully manoeuvring it and swallowing it head-first. Spines and fins, as in the Bullhead (below), are a choking hazard if they are not dealt with properly. Birds may cough up the fish more than once and beat it repeatedly before the meal is finally eaten.

How did it do that?

Another note to *British Birds*, this one from Philip Shooter, relates an incident from 1977. Philip was birdwatching with some friends in Derbyshire, England. They saw a Kingfisher dive from a perch about 3m (10ft) up. When it came up out of the water it had a fish. What makes this unusual is that there was ice over the water, and snow on top of the ice. The bird had blasted through the snow and ice, latched onto a fish and come back out of the same hole it had made. And it had done this even though some of the lake was not iced over. A favourite fishing patch and luck perhaps? While not quite as impressive, it is not unknown for a Kingfisher to choose to fish through an ice hole even when unfrozen water is available nearby.

They do make mistakes

The Kingfisher is not a spear-fisher – it grabs hold of its prey, it doesn't spike it. Not when things go well anyway. Occasionally things go wrong and a fishing Kingfisher surfaces with a fish impaled on its beak, normally on the lower half. The bird kills the fish, but the victim normally goes to waste because the Kingfisher can't manoeuvre an impaled fish into the right position for swallowing. Instead the fish is wrestled off the beak and dropped.

Sometimes a diving bird misses the target fish, and grabs something else instead, perhaps a stone. It can tell the difference though, and normally drops the pseudo-fish pretty quickly. Fishing can also be too effective – the Kingfisher has been known to surface with two fish in its bill. When this happens the bird will whack both fish, get the innermost one into the right position for swallowing, and gulp it down. Unfortunately, while it does this, it loses the outermost fish.

Below and right: They may look impressive but both these dives have gone wrong. Kingfishers normally grab, rather than impale, their catch.

Above: In handling this catch, it was very probably only the innermost fish that the bird was able to hold on to and swallow.

Kingfishers can eat surprisingly large fish, but are occasionally over-ambitious and, after a prolonged battle with their writhing prey, they will admit defeat and let the fish fall. It does not happen often, but sometimes so many fish have been eaten that a Kingfisher has to wait, perhaps for 20 minutes, gripping a fish in its bill, with the fish's head pointing away from the bird, until the bird is ready to swallow yet another fish – swivelling it around so that it goes down head-first.

The winter of 1859–60 was a bitterly cold one in England. In Norfolk, a particularly hungry Kingfisher was creative in its hunting and took something that it really should have avoided. Gulping down its meal proved to be a fatal mistake. This desperate bird had tried to eat a shrew. The account by Rosemary Eastman does not specify which species of shrew the bird had tried to eat. Perhaps it was a Water Shrew (*Neomys fodiens*). This large shrew has poison in its saliva.

Left: Water Shrews are a meal to be avoided, even by desperately hungry Kingfishers.

Special eyes

The Kingfisher has elliptical lenses and excellent eyesight. Good quality optics are worth protecting, and the Kingfisher does this with its 'third eyelid', the nictitating membrane. Just before a plunge-diving bird hits the water, the nictitating membrane covers and protects the eye. When open, this membrane sits at the side of the eye, nearest the bill. When it closes, it crosses the eye horizontally.

When it is fishing, the Kingfisher has to locate fish, select appropriate individuals for capture and consumption, and calculate their underwater position accurately. The fact that the prey is underwater makes the job even harder. Water plays tricks, but the Kingfisher can cope with them. It knows that the fish is deeper down than it looks – otherwise it would misjudge its dives and miss its target. Maybe it also knows that it won't be as big a meal as it looks either. Reflections from the water's surface, especially from a moving surface, can make things even more challenging. The bird keeps its eyes on potential prey, but as it can't move its eyes much, it does this with speedy head movements instead. A human eye has a single fovea, an area on the retina where the image is especially sharp. The Kingfisher has two foveae per eye. Some other birds do too, including hummingbirds, swallows, terns and eagles. It has forward-facing binocular vision and monocular vision to the sides. In each eye one fovea is for the lateral, monocular view and one for the binocular view in front. Should an unfortunate fish appear in the sideways-facing fovea, a small head movement places the fish firmly in the bird's binocular view, where distances can be accurately assessed. The rest is history.

Above: The Kingfisher has excellent eyesight and can fish successfully despite the reflections and distortions that its watery environment presents.

Opposite: When they are underwater, a Kingfisher's eyes are protected by a 'third eyelid'.

Feather care

Above: Time for a preen perhaps.

Feathers maketh the bird and they need to be very well looked after. Kingfisher feathers have to cope with frequent immersion. If a bird is eating 17 fish a day, and has an 80 per cent success rate on its dives, it will put itself through over 20 dives in one day alone. Presumably, digging a nest tunnel takes its toll on the plumage too, and going in and out of one, especially a little later in the season, certainly does.

Feather replacement through moult is vital, as previously discussed. Ongoing maintenance is important too. Preening probably takes up a substantial part of the Kingfisher's day – two hours a day has been suggested. A single session could fill 20 minutes if the job is done properly and could involve bathing a few times as well as the actual preening. During the preen itself the bird uses its bill to apply oil from its preen gland (which is on the bird's rump) to its feathers, to 'zip up' any barbs that have come apart and to make sure any wayward feathers are sitting where they should. The Kingfisher works on most of its plumage with its bill, but there are some bits that are just too tricky to get at – it uses a foot to scratch those hard-to-reach feathers on the head and the back of the neck and will also use the lower surface of a wing to take care of the top of the head. Some wing-stretching completes the routine.

Adult Kingfishers are able to emerge from a dive without looking at all waterlogged. An inexperienced youngster may not have mastered this, however, and the consequences may be terminal. If a young bird struggles to remove itself from the water it will be a little sodden when it finally extricates itself. Ideally, the immature bird needs to get its plumage back into shape before another dive, but it may not do this and the sodden youngster may drown or become too cold to survive.

Above and left: Preening is a vital part of the Kingfisher's day. Most of the body can be reached with the bill but birds will also use their feet for hard-to-reach places (top). A typical preen concludes with a good wing stretch (bottom)..

Finding and Keeping a Territory

How territorial is the Kingfisher? In a word – very. Don't be drawn in by its gorgeous colours – the Kingfisher is a feisty creature, and many defend breeding and winter feeding territories. Most birds spend more than half of the year living a solitary existence. Breeding takes two, of course, but as the breeding season comes to a close, the male and female go their separate ways. Birds that winter around the coast seem to be less territorial, and are more tolerant of their fellow Kingfishers.

By early autumn, winter feeding territories are being secured. An adult male's winter domain is usually the same patch as his summer breeding territory. Photographer Ian Llewellyn watched a male in Avon, south-west England, that held a territory continuously for at least three breeding seasons and two winters. The female may hold a winter territory that abuts that of the male she bred with, or at

Opposite and below: Kingfishers are very territorial and may defend their patch with considerable aggression. If threat postures fail to intimidate, the birds will launch into battle, using bills as weapons.

Right: A Grey Wagtail may be driven off by a territorial Kingfisher.

least, is not too far away. Sometimes, the breeding territory is split between the male and female to form two winter feeding territories. These may overlap to some extent, but this is quite rare. David Boag watched a male and female that had bred together go on to share an artificially stocked fishing pool (effectively a bird table for Kingfishers) through a winter. They shared the easy pickings, but wouldn't fish there at the same time, even though they had previously fished in close proximity.

The Kingfisher's aggression is not only directed at other Kingfishers. An important perch in a bird's territory will not be shared lightly and the blue blur may chase impostors off – the ousting of both a Grey Wagtail (*Motacilla cinerea*) and a European Robin (*Erithacus rubecula*) have been recorded.

Juveniles begin their lives in the territory that they hatched in, but their parents do not tolerate their presence there for long. For a while, they get along just fine with their fellow offspring, but there comes a point when the desire to 'own property' kicks in, and sibling relationships deteriorate.

This is a tricky time for the newly independent bird, and juvenile territory hunters turn up in surprising places, including garden ponds. At first, the youngster avoids advertising its presence. Kingfishers often give

away their presence by their call, but the juveniles are smart enough to stay silent when they are new to an area. If they are on another bird's patch, and they are found, the territory owner will chase them off. But if no existing 'property owner' shows up, with time the incomer gets more vocal, though a little tentatively at first. Within a couple of days, the young bird could be shouting out its presence and ownership with full gusto calls as it flies around its new home. Birds that fail to secure their own winter feeding territory may be among those that move to coastal waters for the winter.

Above: Aggressive behaviour between two young Kingfishers.

Below: Juvenile Kingfishers, like this one, need to find their own territory.

How big is a territory?

Above: The straight back, beady stare and open bill show that this male Kingfisher is ready for a challenge. An upright threat posture such as this may be followed by more dramatic behaviour.

Ian Llewellyn's observations in Avon, England, suggest that there can be two or three nests on a 1.6km (1 mile) stretch of river. This is an average of one nest every 500m (1,640ft) or so. Pairs tend to keep their distance from each other. On an Estonian river, the gap between pairs ranged from 300m (980ft) to 1km (²/₃ mile), with as many as 15 pairs over 18km (11 miles), while gaps of just 125m (410ft), and 150m (490ft) have been recorded in Sweden and Scotland respectively. There have been plenty of territories recorded that occupy as much as 3km (1¾ miles) or even 5km (3 miles) and a Belgian male had a home range that was nearly 14km (8²/₃ miles) long. With territories of this size, some parts are seldom visited, but visitors, if discovered, will not be welcomed. Rosemary Eastman's work found that a pair would each have their own fishing patch in the breeding territory, but would not be above poaching a fish from their partner's stock. Clearly, territory size varies, but when one pair is just too close to another, stress levels are high and productivity suffers. The areas defended in the winter tend to be smaller than those defended in the summer, and during particularly cold weather more birds than normal may feed where food is in good supply.

A study in Estonia by Kumari, published back in 1939, concluded that a territory includes a core area that is particularly important to the male, and when birds gather food for their offspring, they fish in a feeding territory that is apart from the breeding territory, though usually within a kilometre (²/₃ mile) of the nest. How this relates to other wisdom on Kingfisher territories is unclear.

What happens when rivals meet?

A Kingfisher will defend its patch with considerable zeal, be it a breeding territory or a winter feeding territory. In-flight calls, and calls prior to landing, are the bird's declaration of ownership, but these don't keep out all unwelcome visitors and it is not unusual for invaders to show up. In a breeding territory both adults play a part in challenging other Kingfishers that dare to enter their patch, and will take them on either alone, or as a double act.

If an invader is discovered, there's a roughly 50:50 chance that evicting it will need nothing more than a

Below: This female Kingfisher is in threat mode.

determined chase. But it is not always that easy. If the impostor is made of tougher stuff, things escalate, with the resident and the impostor beginning threat displays. When this happens, the gap between the two rivals is often about a metre (3¼ft). They stand tall with their feathers flattened and their shoulders up. Their wings sometimes sag forward and their eyes are firmly on their opponent. The bill is agape, a little elevated or pointing forwards on an elongated neck, and swinging unhurriedly from side to side, but not pointing at the 'enemy'. In the silence, the tension is palpable, but it dissipates if the display moves into aerial pursuit.

What then?

The upright threat display can be a prelude to a forward threat display, where one or both birds gradually bends forward until it is nearer horizontal than vertical, perhaps lying, again with elongated neck, and sometimes swinging slowly from side to side. There could be jerky movements and calls, or no vocal accompaniment at all. By presenting a sideways view, threat displays may be a show of strength, enabling participants to show off the size of their bill. Opponents have been seen doing synchronised place swaps, and shifting slowly from one threat type to another, and back again. The threat repertoire also includes a stooped, spread-winged posture, and bobbing the tail.

Below: The low threat posture, seen to full effect in this juvenile, follows the upright threat posture. If neither opponent backs down, such aggressive posturing may lead to fights, especially when birds meet at territorial boundaries.

Above: Two adult Kingfishers in combat. Their bills can be used aggressively as well as for feeding, preening and tunnel excavation.

If threat displays fail to oust an intruder, things can get very nasty and very noisy. A stream of varied, harsh, rasping calls breaks any silence that remained and the attacker takes to the air, intent on forcing its enemy off its perch. Bills are used as weapons. The birds may tumble down, and a watery wrestling match might ensue, each trying to get the other's head underwater, perhaps even attempting to drown the enemy.

Typically, battles do not last long, but there are records of fights carrying on for many hours. Ian Llewellyn has seen males posturing to each other outside a nest, with stretched necks and pointing beaks, for three hours. Neither bird conceded and the birds then fought physically, beak to beak and beak to wing. Ian has observed that when fights occur, they are often at nest banks or a territorial boundary. In 2009 he observed ongoing conflict between four birds near a nest hole in Bristol. The encounter involved breeding birds that were determined rivals and a completed nest site – a recipe for war. The birds refused to back down and tensions built over three or four days, with huge aerial chases, high-pitched, staccato calls, stabbing attacks, attempts to

knock each other off perches and plenty of exaggerated posturing. Sometimes, the males would take time out from the fracas and retire to their nest tunnel, leaving the females to battle it out. When these encounters are at their most intense, the combatants are so absorbed in conflict that they rarely fish and by the end of the day, they are panting and very tired. When neither male admitted defeat the battle ended with the females fighting in the water, with locked beaks, for five or six seconds. After that things calmed down.

David Boag suggested that the males get involved in more disputes than the females, but Ian Llewellyn has seen more females fighting and is very clear that 'the most intense, sustained, aggressive and absolute' are the females.

There are also stories of aggressive encounters between two Kingfishers which are observed by a third, non-participating bird.

In the early part of the breeding season tempers can be very friable. Things can get particularly intense if pairs attempt to nest near their neighbours. In these situations a rival's hopes are sometimes foiled by a visit to their nest tunnel. This is not a social visit – the intruder is there to pierce the eggs. An account from Scotland tells of two nests that were close together with an unused tunnel separating them, that both pairs had rejected. A third pair adopted this nest, and laid seven eggs in it, but the eggs didn't survive long. The adults returned one day to find that their neighbours, who had young birds in their nests, had evicted the eggs.

Opposite: Kingfisher aggression can include aerial attacks, with the intent of forcing a bird to abandon its perch.

69

FINDING AND KEEPING A TERRITORY

Tired and confused?

To some people's ears, the call of the Dunnock (*Prunella modularis*) bears an uncanny resemblance to one of the Kingfisher's. It has even been known to fool a Kingfisher. David Boag recounts a tale of a Kingfisher, that had only just seen off an invader, which heard a Dunnock call and went straight back into threat mode.

Dunnock.

A Mate and a Tunnel

After time spent alone the transition to life with a partner can be tricky. The Kingfishers that have survived the winter, including those that are less than one calendar year old, need to find a mate. By February, and sometimes earlier, Kingfishers in southern England can be looking for a partner. This may start a little later further north.

After the relative quiet of the winter months, February and March can be one of the best times of the year to see and hear Kingfishers. The quest for a mate can begin quite sedately, with a perched bird declaring his presence with intermittent calls. A study by Heyn, published in 1963, describes a bird that kept this up for an hour before a potential mate made an appearance. The closer the new arrival got, the more the first bird called, until perches were abandoned in favour of an aerial chase.

Opposite: When a male Kingfisher presents a female with a fish as part of their courtship, he passes it to her head-first.

Below: A Kingfisher and its mate with their nest tunnel in the background. The bird in flight appears to have soil on its beak, so may well have been excavating the tunnel.

Aerial wooing

Their high-octane courtship flights are noisy aerial pursuits that can take the two birds from just above the water to the tops of the trees in an instant, where they might settle, with high-pitched whistles drawing attention to their antics. Some energy-sapping chases last for hours, and from an observer's point of view, it can be hard to work out if you are watching aggression or romance. The reality is that the chase could include elements of both. Normally it is just two birds that take part in a courtship flight, but not always – five or more participants have been recorded. Courtship flights are at their most vigorous before the first brood of the season, but can be seen later in the year too, when they are performed, albeit with rather less zeal, as a prelude to the next brood.

More exotic courtship flights include approaching the female, perhaps more than once, with an elegant glide as the finale to a semicircular flight that took the male out over dry land. Gliding is not a Kingfisher's normal flight mode. Both genders have also been seen flying gently on quivering wings, low over the water, and there is a record from *British Birds* in 1927 (contributed by W. M. Marsden) of non-stop 'singing' from two Kingfishers as they traced tentative

Below: This male is waiting with a fishy gift for the incoming female. Or, given that the head of the fish is facing towards the male's beak, maybe he will eat it himself.

circular flight paths 15–20 metres (*c.* 50–65ft) in diameter, with sporadic pauses in nearby bushes. A similarly vocal male was seen flying rings around a less energetic female, who remained perched (A. van Beneden, 1930). The 'song' of the birds in question was described as similar to that of a European Greenfinch (*Chloris chloris*) or Dipper (*Cinclus cinclus*), but not all Kingfisher aficionados are convinced that it really was that complex, and suggest that it was probably a simple series of differing whistles.

Love or hate?

Above: A male adopting a threat posture during courtship. This is not unusual early in the breeding season, when males are not at ease in the presence of a female.

At the beginning of the breeding season, a male Kingfisher can be quite uncomfortable in the company of a female and might, at first, display aggressively towards her. With time, things change. Both genders take part in the courtship display, but the males do it more, and they ham it up more as well.

The courtship display is not dissimilar to the upright threat display. A displaying bird adopts a very vertical posture with its head pointing a little above horizontal and its wings draped forward so that the wingtips are as low as the hopeful performer's feet. Once again, the

Kingfisher positions itself in a way that makes the size of its bill obvious. And if this were not enough to impress a potential partner, there is sometimes some soft whistling by way of vocal accompaniment. Courting birds take it in turns to whistle, and do not make this sound at any other time. As things progress, the gap between whistles reduces, so that ultimately, the illusion of a single note is created. Within a few days of pairing up, presumably as the pair bond becomes more secure, courtship displays become less frequent.

A gift of fish

It may not work in our society, but it does the job for the Kingfisher. When the tentative ambiguity of early courtship is over, the male Kingfisher presents his partner with the gift of a fish and, for a while, this could take place a few times every day. The gift is sometimes a sequel to an aerial chase after which the male leaves the female perched and goes fishing. When a male catches a fish intended for his partner, he swivels it around so that its head points out, contrary to the normal procedure when the fish is for his own consumption and the head must point in for safe swallowing. As he returns there are animated contact calls from both parties – the male can call without letting go of his fishy gift. The female might now move closer to the male or do a bit of courtship displaying while he stoops, with wings lowered, and reaches towards her with his fishy offering.

The female quivers her wings as she takes his gift and the male might complete the ritual with a bit of courtship display. As an alternative, and to quote Ian Llewellyn, the male, with bill pointing to the sky, 'goes bolt upright, fans his tail feathers and loops and dives like an Olympic diver, or loops and goes back onto the perch'.

Another description involves the male, laden with a fishy gift, calling in flight (again, without dropping the fish) and using the female's response to pinpoint her position. Once perched, a subsequent, very short flight from either party and a bit of sidling along the perch brings the pair

Above and below: Once early courtship is complete, male Kingfishers will present their partner with fish, perhaps several times a day.

even closer. Sometimes hungry females will beg for food from their partner and, in contrast to the grand fish-passing event described above, there are occasions when a female simply grabs the fish and gulps it down.

Is monogamy the only way?

For the Kingfisher, monogamy is not the only way. Mostly they are monogamous, but not necessarily with the same partner from one year to the next. Where pairs are constant from year to year, and this is not unusual, it is likely to be two birds that carved up their breeding territory into two winter territories, with one territory for each bird.

It is also not unusual for a bird to change its mate from one year to the next, if, for example, an incomer has taken over their previous mate's winter territory.

Polygamy is not the norm, but it is far from unheard of, with a typical polygamous relationship connecting two females to one male. Different studies have produced different figures for the incidence of polygamy, ranging from 6 per cent in a Swedish study by S. Svensson that looked at 116 'nesting associations' to about 35 per cent on the Oka reserve in Russia (Numerov and Kotyukov, 1979). Occasionally, the polygamy involves three females, and in one of the Russian examples above, because the females were double-brooded, the male had his work cut out helping at six nests. Two females is more normal. The male may initiate proceedings with a second partner when the first is busy looking after eggs or brooding chicks. He might even abandon his first partner, leaving her with all of the hard work of raising the offspring. Sometimes both partners are at the same stage and the male incubates at one or both of the nests, leaving the remainder to the female.

Ian Llewellyn saw polygamy in action when a male mated with a second female just before his chicks with the first female fledged. He continued to take food to the first nest, but when his first partner ventured near the nest site of the second partner she was attacked by both the female and the male, with the females fighting throughout the day.

Digging a hole for yourself

The Kingfisher's nest is a chamber at the end of a tunnel, and, soon after securing a partner, the search for an appropriate location is a priority. A pair either uses an existing tunnel or digs a new one, and after an energetic aerial chase, courting birds take a look at 'used homes' or begin to create one of their own. A 'used home' may be considered, and even worked on, before being rejected in favour of building from scratch. At this stage the male normally does the work. Studies found that most nest sites are being excavated in March, though work can take place in February and April too. Birds that take longer to find a partner do not have to contend with the vagaries of the weather to the same degree that earlier paired birds do. They court less before building, which starts sooner in the relationship, and seem to put more effort into excavating. As a consequence, late paired birds will normally be on eggs no more than a few weeks later than their more precocious neighbours.

Below: The ideal nest site has soft sandy soil that is easy to excavate. Male and female Kingfishers share the work of making the tunnel by pecking repeatedly at the bank until a foothold has been created and then taking turns to dig.

Right: A 'blue blur' streaking into its tunnel on a riverbank. Such a low nesting site could be risky for this Kingfisher and its family, increasing the threat from flooding or predation.

Some birds go on to a second brood, and a few add a third. Often these are in the 'original' chamber, but sometimes a new nest is created. This can be nearby or quite a distance away – one was 5.5km (3¹/₃ miles) from the original nest site, and this was for a brood that started before the first one had fledged.

An ideal nest site is easy to excavate and safe from predators and flooding. Nest tunnels are found in vertical, steep or overhanging banks on the edges of streams, rivers, canals, ditches and lakes, with no preference for banks facing in a particular direction.

Soft, sandy soils with no roots or stones make digging straightforward and being off the ground in a bank makes access trickier for predators, especially if there are no plants

Left and below: Once a foothold has been created the Kingfishers can start digging in earnest, using their bills to excavate a nest tunnel. The dimensions of the tunnel (as seen below looking out from the inside) are such that the birds cannot turn around until they have excavated or at least begun work on the chamber; instead they must walk backwards to exit, pushing loose soil behind them with their tails as they do so.

around the nest hole. However, plants can also help to hide the nest from animals one step up the food chain. Typically, the tunnel entrance is 90–180cm (3–6ft) up the bank, which affords a certain amount of protection from rising waters. There is an old record of a nest that was 36m (120ft) above the water's surface – presumably there was no danger of that nest flooding! Most tunnels are 50cm (1²/₃ft) or less below the top of the bank. Occasionally, Kingfishers will excavate a tunnel that is not by water, but this is unusual.

Some sites are used year after year, even when there are good alternatives in the area. Over a four-year period, Ian Llewellyn watched a poor quality Kingfisher bank that the birds persisted in using despite the fact that the nests were consistently either flooded or predated.

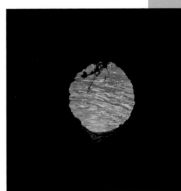

Other nest sites

Not all Kingfisher nests are in a 'standard' waterside bank. Some use the soil around the roots of a toppled tree. This could be more or less vertical and near the water, but if they hit a root they have a problem and may well give up. Others use banks that are well away from the water – David Boag's book described a pair that bred in an earth bank in a patch of woodland. Other sub-prime sites include banks on minor country roads, nooks and crannies in caves, bridges, walls and the banks of railway cuttings.

Kingfishers are reported to very occasionally take advantage of another animal's labours, moving in to tunnels excavated by a Bank Vole (*Myodes glareolus*), Sand Martin (*Riparia riparia*), or more surprisingly, a Rabbit (*Oryctolagus cuniculus*).

It is not unknown for Kingfishers to use a Sand Martin tunnel as their nest site.

And sometimes our own species gives a helping hand, deliberately creating artificial nest sites to encourage Kingfishers to breed. If the location and design are good the birds can be quick to move in.

Design and dimensions

The tunnel leading to the nesting chamber is usually around 45–90cm (1½–3ft) long, and either has a gentle incline towards the chamber or is horizontal. An incline aids the drainage of bird poo, which is an important function. Some tunnels are shorter and some are longer, depending, at least in part, on what the birds have had to dig through – a tunnel of just 15cm (6in) and another of about 135cm (4½ft) have been recorded. The best tunnels are straight ones, as these are the most effective at allowing daylight into the chamber.

Studies have shown that Kingfishers are very picky about the diameter of the nest tunnel, digging a route in that is more or less 5.5cm (2¼in) across. Tunnels have also been recorded with a diameter of 6–7cm (2⅓–2¾in).

The tunnel ends in a rounded chamber where the eggs are laid and the young looked after. The size of the chamber varies, but on average is about 11cm (4⅓in) from top to bottom, and around 16 or 17cm (6⅓–6⅔in) from left to right and front to back. The birds create a dip in the base of the chamber to increase the chances of their precious eggs staying put. They do not line the nest.

Excavating the nest tunnel is a shared task, with both male and female playing a part, though the male may do most of the work. The time taken depends, among other things, on how hard the soil is to dig and what obstacles are met on the way – stones or roots slow down progress and can force the project to be abandoned. When preparing the first nest of the season, the female might contribute almost as much as the male but if a second tunnel is excavated the male may do virtually everything. Most digging is done in the mornings and 7–12 days normally sees the job completed. Early paired birds can be a bit more leisurely, taking as long as two weeks to finish the project, while hard-working late paired Kingfishers can get the job done in as little as three days. If a second nest is excavated in the same season less time tends to be spent on it – usually 4–7 days.

Contrary to some established wisdom, Ian Llewellyn found that a male will sometimes check out existing tunnels and do a bit of digging alone, before he has found a partner. He also believes that birds investigate nest sites at any time of the year, though with increasing curiosity as spring approaches, and has seen a male with fresh mud on its bill in mid-November.

Below: It's artificial, but may well be fit for purpose!

The technique

Excavation begins with the strange spectacle of a Kingfisher flying at a bank and pecking it. The male starts the process, flying from his perch without warning and pecking the chosen bank, with his partner nearby. He comes back to his perch and shakes or scrapes the soil off his bill, but may fly and peck repeatedly to secure the female's active involvement. When she does get involved both birds launch themselves towards their potential new home, pecking alternately at different spots on the bank. There comes a time when the apparent randomness disappears and all their efforts focus on one place. The aim is to create a foothold from which the Kingfisher does a passable impression of a woodpecker as excavation continues, one bird on, one bird off, each taking their turn. The 'excavator' can hear the muted whistles of its partner, who rests and acts as lookout. Should trouble appear the lookout's call changes and the construction site is quickly abandoned. Ian Llewellyn witnessed a digging operation that was abandoned

Below, left to right: Male and female partners take turns to excavate their net. The site here is less than ideal for this pair; it has lots of roots to contend with and heavy mud.

for good because of the presence of a Sparrowhawk (*Accipiter nisus*) in the area.

As the tunnel gets longer, things get easier unless, of course, they hit an obstacle. Digging birds use their feet to push the soil behind them as they progress. The dimensions of the lengthening tunnel are such that a tunnelling bird cannot turn around and has to walk backwards to extricate itself. As it does so, its tail turns the bird from excavator to mini-bulldozer and loose soil is pushed back down the tunnel.

Digging a tunnel is a bit of a hit and miss affair – hopefully, the chosen route will miss impenetrable obstacles, but sometimes of course it hits them, and this could happen more than once before the birds choose a good spot. When it happens, one way forward is to dig around the object and try to remove it. If this is unsuccessful there are several options. The bird could complete the nest site with a bend in the tunnel, or, if it is deep enough in the bank, it could stop where it is and create the chamber by the barrier. Some birds elect to do neither of these, give up on the tunnel and start another one.

Finished

If you have the privilege of watching Kingfishers digging out their nest, there is an easy way of knowing when work on the chamber is at least underway. As previously discussed (see page 83), the narrow tunnel forces a tunnelling bird to come out backwards because there is no room to turn around. When the bird comes out beak first more space must have been created inside so work on the chamber must have begun.

Studies indicate that mating normally takes place between mid-March and mid-April, when nest excavation is finished or nearly finished. Often the gift of a fish comes before the act itself, but not always. Ian Llewellyn has seen Kingfishers mating 9–12m (30–40ft) up in a tree, though it usually takes place on a branch that is near the nest site and about 1.2–1.8m (4–6ft) high. The male's advances are not always welcomed, and an unwilling female makes her uncooperative state plain with her body language and a voice that has been described as both 'soft hissing' and 'angry'. Copulation starts with the birds alongside each other, and the male then hovering to get into position, or with the male perching above his mate and dropping down from there. A willing female may

Below and opposite: Mating often takes place on a branch close to the nest site. Kingfishers do not usually mate until nest excavation is almost complete.

adopt a position similar to the head-up courtship posture
described earlier (see page 73). The female will quiver
her sagging wings and urge the male on with encouraging
noises while he hovers above her. The female will then
flatten her body and the male's feet will grip her back
while his bill holds on to her crown feathers. Male
Kingfishers can be quite rough, and may pull the female's
head around as they flap to maintain their balance. The
female will move her tail out of the way then the male's
cloaca will touch hers. The contact can be short-lived –
as little as seven seconds or so – and genes for a future
generation pass quickly from male to female. When the
act is complete the male normally flies away, and perhaps
bathes. David Boag observed a female fly off ten minutes
later while Rosemary Eastman's account tells of a post-
coital female that hung around for 15 minutes, at which
point the male brought her a fish and she retired to the
nest tunnel.

Mating may be a less elaborate affair prior to a second
nesting attempt, with much less preamble and no active
encouragement from the female. Birds also mate less
often second time round.

From Egg to Adult

All being well, mating means eggs, eggs turn into nestlings, and nestlings become fledglings. But the adult birds have a lot of work to do along the way before encouraging their offspring to leave the parental home.

One nest chamber can see a pair of Kingfishers through the entire breeding season, even if they have more than one brood, though plenty of birds move to a different tunnel (or tunnels) after brood number one. When broods overlap, a second tunnel is essential. If the pair goes as far as a third brood, it may return to its first nest.

Opposite: Kingfishers don't start their life looking like this!

Left: It is hard to believe that these newly hatched chicks will transform into one of our best-loved birds. A clutch of six or seven eggs is typical, and it is not uncommon for Kingfishers to have up to three broods in a season. Regurgitated fish bones and scales can be clearly seen in this picture.

Fish-bone nests

In Greek mythology the Kingfisher's fragile, floating nest was said to have been made of fish bones. The truth is that Kingfishers do not build a nest of fish bones. They do not collect any material to deliberately line their nests but eating fish in a confined space means that with time, a layer of regurgitated fish bones and scales builds up, supplemented by the shed quill cases of the fledged youngsters. This is the 'nest'. Keeping the resulting 'fish bone cup' in one piece is very tricky, but apparently there was a time when anyone who could find one and keep it intact would be very well rewarded, with the King of England offering a bag of gold and the British Museum £100 for one.

Eggs and broods

Above: A clutch of Kingfisher eggs.

A typical clutch consists of six or seven smooth, shiny, round, white eggs. Clutches as small as four are possible, and sometimes eight eggs are laid. One egg weighs just over 4g (⅛oz) and is about 23mm (just under 1in) long and 19mm (¾in) wide. An egg is laid each day until the clutch is complete then incubation begins. The eggs are described as initially a delicate pink, which then turned white.

Kingfishers are usually single- or double-brooded, so, with clutches of 6–7 eggs, could produce 6–14 young in one season. David Boag found a three-brood pair that had lost their first brood to flooding and whose second and third broods overlapped. Domestic duties were shared, with the female taking care of the new eggs and the male ferrying food to the growing youngsters of their second brood. One male who kept two females was known to be triple-brooded twice over. That makes six broods, so if each of these had just six eggs in it, he would have sired 36 offspring that season.

Incubation

Both parents incubate and textbook wisdom is that the task is divided up more or less equally. Studies have found that the females are a bit keener to incubate than the males: the Eastmans recorded males sitting for about three quarters of an hour, and the female for about an hour and a quarter, and a study by Zoller gives a sitting time of two and a half to four and a half hours.

Changeover can be a relatively simple affair. The absent parent flies towards the nest site, calling a few times, and settles nearby. The incubating bird comes out, and the other parent goes in. More elaborate changeovers also occur, with a bit of courtship display and food being delivered to the previously tunnel-bound bird. Eggs are not normally left unattended though sometimes they are in the early stages when the birds are still working out the routine. The off-duty bird goes fishing, while the on-duty bird finds its way down the tunnel, moves to the back of the eggs and carefully

manoeuvres into position. It uses its bill to move the
eggs beneath its body, conveniently turning them in
the process. Kingfishers do not seem to be the most
observant or diligent of parents – an egg that has become
separated from the others may be left to its own devices,
with bad consequences. The sitting bird faces the tunnel
entrance and might begin its stint with a relaxing preen.
At some point, a pellet will probably be coughed up –
the nest becomes increasingly unpleasant as things
move on.

Above: This vibrant female has
just emerged from the nest after
spending time incubating.

Below: A female in the nest
chamber with a clutch of
seven eggs.

Right: The flash of blue says it all – that hole is a Kingfisher nest tunnel.

The Eastmans watched a second-brood nest and found that the female took responsibility for overnight incubation. The male presumably spends the night at a safe roost site, tucked away in a waterside tree or bush, which could be a fair way from his eggs and partner.

Task allocation varies when a later brood overlaps an earlier one, which is not unusual. Males have been recorded incubating later clutches and leaving the female to look after the young of an earlier clutch. There is even a record of a hard-working male taking food to one set of offspring while doing a bit of incubation elsewhere. Both birds normally care for the last clutch of the year.

However the work is carved up, incubation takes about three weeks.

Above and below: A male approaching the nest with food. And then departing.

Hatching

Because Kingfishers do not start incubating until all the eggs are laid, the young emerge at a similar time, with just a few hours between each hatching. The eggs have done their job and an adult will diligently take each broken eggshell out of the nest, airlifting it in its bill to take it away from the tunnel, and letting it fall into the water below. In some ways this is a strange thing to do, as the tunnel and chamber are certainly not a model of hygienic living – with excrement, undigested fish bones and scales creating a 'special ambience'. Birds that build nests that are more accessible than the Kingfisher's make their nest and young less obvious to predators by removing the eggshells, but this is unlikely to be an issue for the

Right: When Kingfishers hatch they are pink and blind.

Kingfisher. Scientists suggest that another reason to remove eggshells might be to make sure that the later hatching eggs in a clutch aren't impeded by empty eggshells.

The young Kingfishers are pink, blind, have no feathers and are totally dependent on their parents. They are brooded by one parent or the other for the first six to ten days and are fed their first fish dinner when they are just a few hours old.

Above: The male and female take it in turns to brood their young family.

Below: These tiny Kingfishers need to be fed!

Fish, fish and more fish

Above: Chicks are fed whole fish by their parents. Initially, the adults choose smaller fish when catching a meal for their young. This male is about to enter the nest with a meal for one of his brood.

It has been estimated that a typical nest of Kingfishers will, at its peak, get through more than 100 fish a day, and that is just the chicks. For the first week and a half or so, the adults bring relatively small fish in to feed the youngsters, limiting their catch to fish no more than 5cm (2in) long. As they make their way down the tunnel, they block out the light and, in response to the darkness, the chicks raise their heads. A chick gets a whole fish, and sometimes a surprisingly large one, taking it head-first. Even at this stage, both mandibles are tipped with a flash of white, a distinctive feature retained on fledged juvenile birds. The white reflects what light there is, and may give the parent bird something to aim at. Getting the fish down is not always easy – David Boag saw a young Kingfisher wait for ten minutes with the tail end sticking out of its bill. Their guts must be very effective – in a matter of minutes digestive juices are breaking down the fish's head and making room for the rest of the meal.

The chicks are silent at first, but as their size increases so does the noise level. When they are about a week and a half old the youngsters get much noisier, making a continual churring noise that can be heard as far as 30m (100ft) away. It is probably made by just one bird at a time – the one next in line for a fish – though two birds may call at once. If the bird is not hungry the churring ceases, but soon starts up again when an adult comes back to the nest. The bigger the chicks get, the louder the noise becomes, but it stops if the birds sense danger.

By the time the chicks are a week old, the demand for fresh fish deliveries has risen. As a fish-laden adult flies home to an expectant brood, it might utter a *teee ti-tee ti-tee ti-tee* contact call and as it arrives at the nest the churring from within gets more animated. When the chick at the front of the brood has been fed it moves around to the back. In its eagerness to be fed, the young bird sometimes grabs hold of the adult's bill as well as the fish and a bit of a tussle takes place. But the chicks keep themselves in order, and the fed bird goes to the back of the crowd. When the next fish arrives the next chick is fed, and so it goes on, with the fish going down head-first of course. As much as 50 minutes can pass before a chick gets its next fish and if it dares to go for more than its fair share, its siblings will give it a good pecking.

Left: By the time they are 12 days old the chicks are very vocal and will call when they are hungry. The continual churring noise can be heard from 30m away.

Growing up

At seven days old, the chicks are more robust and much of the characteristic blue colouring is visible, albeit in muted tones, as feathers that have yet to emerge become visible through the skin. It may take five or six more days for the feathers to poke through the skin. When they do, they are encased in sheaths, and bit by bit, the nestling is transformed into a mass of dull blue and orange spikes. As the sheaths break open the feathers appear, tip first, but this does not start until the birds are about three weeks old and is more or less finished a day or two before the birds fledge. The eyes begin to open at eight days old, or a little later.

As the chicks get older the parents are more likely to reverse out of the tunnel than to come out head-first, and when the chicks are two or three weeks old they organise themselves differently. They are more and more likely to line up side by side in rows, with their bills pointing towards the tunnel. By this age a hungry youngster will quiver its wings as it requests yet more food, and as the time to abandon their subterranean existence draws near, most of the dining takes place in the tunnel, rather than the chamber, and waiting in turn to eat tends to become a thing of the past.

Right: These 16-day-old chicks have lined up in rows facing the tunnel entrance. Their feathers are encased in sheaths until the birds are about three weeks old, giving the impression that they are covered in spikes.

Mostly, young Kingfishers eat fish, but not always. Ian Llewellyn watched a bird in Avon bring Signal Crayfish (*Pacifastacus leniusculus*) to the nest to feed the young. He saw ten visits and crayfish were the dish of the day on five of them. The crayfish were about the size of a man's thumb, with their big front claws missing, so presumably the bird had worked out how to de-claw them. At the time the river was brown with mud, which must have made fishing trickier than normal. Adults have been known to eat crayfish, but this may be the first time that crayfish have been recorded as food for nestlings.

Above: Adults will continue to feed their young for a short while after they have fledged. This youngster begging for food will soon have to fend for itself.

Below: Signal Crayfish – not all young Kingfishers are fed entirely on fish.

Keeping it clean

The chamber reeks of regurgitated fish remains from the youngsters and their parents, but the young birds aren't keen to sit in their own excrement. They see the light from the tunnel entrance, and to make life in the chamber more tolerable, they squirt their white stuff towards the tunnel, like a blast from a high-pressure hose. This helps to keep the chamber clean but turns the tunnel into a slimy, excreta-coated mess that the parents have to traverse many times a day, making frequent

Above and below: Thanks to the mess the chicks make, going in and out of the tunnel is a messy business – bathing helps to keep the adults' plumage in good condition.

baths a necessity. And as the youngsters near fledging, their discarded feather sheaths are added to the previous weeks' leftovers. When the young have fledged, the tunnel and chamber could be home to another brood, in which case the female might clear out the bones before taking matters any further.

Into the light

The nest is a quieter place just before fledging. Instead of calling the youngsters stretch their wings and preen their recently acquired feathers. There are limits though, and their vow of silence is conveniently forgotten if food is at hand, when they give a rough *gred gred* call. The parents may be quiet too, and bring less and less food to their offspring. The departure itself can take place with no ceremony – a juvenile just flies out, and then another one, sometimes without a parent in attendance. The parents are there sometimes though, and studies by Heyn (1963) and Reinsch (1962) talk of the adults calling to urge the young on and out. Boag's work, however, saw nothing like this. Even if the bird is around it is not always the model of a good parent – the Eastmans watched a female that remained a detached observer and then left the area when one of her brood dropped into the water after just a few flaps, and was then rescued by Ron and Rosemary.

Above: Note the pale bill-tip and dark breast feathers on this young Kingfisher.

Generally all the young leave the nest on one day, with gaps between departures of 10 to 20 minutes, though this can extend into a second day. There is no going back, and those that do not fall into the water head for a nearby perch or are taken fishing by one of their parents.

Kingfishers fledge after 23–27 days in the nest. Fledging rates vary – Clancey's 1935 study showed that 80 per cent of the eggs that hatched made it to fledging; while Glutz and Bauer's 1980 publication gives a lower figure of 53.8 per cent for Swiss Kingfishers, but that was based on the number of eggs that were laid rather than the number that hatched.

Independence day

Juvenile Kingfishers move away from the nest site fairly swiftly. Within a day they can be 300m (980ft) away and given a couple more days can be 4km (2½ miles) away.

The adults will continue to feed the young when they are out of the nest, but not for long. To help its parent find it, a youngster gives a *chip* or *chip chip* call, a note that it has used since day 17 or 18 in the nest. For a youngster to survive it must hone its fishing skills as soon as possible and will be diving on the same day that it fledges. Many

Below: Two young siblings eager for a meal! The bird in the middle is one of the parents – the bright orange feet are a feature of an adult Kingfisher.

Left: This very young Kingfisher has fledged, but must sharpen its fishing skills to get through the winter. Its chances of survival are not good.

flounder and become waterlogged; some drown, others make it to the bank only to perish there, while others dry out and live to dive another dive. Some do well – one was taking pond-skaters (*Gerris* sp.) just one hour after leaving the nest.

Whatever the state of their fishing skills, the adults want the youngsters off their patch within four days of them leaving the nest and sometimes sooner. They will 'encourage' them to leave, by chasing and calling, if that is what it takes. One study saw an adult switch from feeding one of her offspring one minute, to driving another one out of the territory, and then telling the one she had been feeding to do likewise. Such oustings are accompanied by angry calls, making Kingfisher country pretty noisy when there are young to move on.

The evicted youngsters must look for their own territory, with no knowledge of who lives where. If they are fishing on another bird's patch that bird will want to move them on, but the incomer keeps quiet and if its presence goes unchallenged the new Kingfisher on the block will stake its claim.

Life and Death on the Riverbank

According to data on the British Trust for Ornithology's website a little more than 20 per cent of the Kingfishers that fledge survive their first year – four out of five do not make it to their first birthday. This may sound grim but many bird species have similar mortality rates during this part of their life cycle. For the one in five that survive long enough to have a go at breeding, the odds improve, but not by much: every year more than 70 per cent perish and a Kingfisher's typical life expectancy is only two years, though some live much longer.

The British record is held by a Kingfisher that lived in Hampshire and made it to the grand old age of four and a half years. It is likely there are longer-lived British birds – a Belgian Kingfisher survived for over an astonishing 21 years.

Lethal weather

Particularly cold winters can have devastating effects on Kingfisher populations. When feeding areas freeze over, apart from any intrepid individuals who go in through the ice, hungry birds must avoid starvation by finding their food elsewhere, taking different prey or moving to a different area. In the freeze of 1889 dead Kingfishers were being 'picked up daily' around the River Kennet in the south of England. The winter of 1939–40 was so severe that the Thames iced over. Five years previously, a 109km (68 mile) stretch of the river was home to around 120 Kingfishers. After the freeze just a pair or two remained. In Britain, the icy winter of 1962–63 killed birds of many species. The blue blur had a very tough time – ornithologists think that 85–90 per cent were lost.

Opposite: A handsome male Kingfisher – but life on the riverbank can be tough.

Top: Many Kingfishers perish during icy winters and, as a result of climate change, these conditions are likely to become more frequent. How Kingfishers will cope remains to be seen.

Above: Floods can destroy Kingfisher breeding sites and make it harder for them to fish.

The cold of that winter was such that even estuaries iced over, so birds that moved to the coast would have struggled too. Cold winters in 1978–79 and 1981–82 also made an impact on the UK population, but numbers rose between the mid-1980s and in the years up to 2007. Breeding Bird Survey data show that after the severe winter weather of 2009–10 numbers dropped.

The following winter was another brutal one, and breeding numbers in 2011 were even lower. Since then, the Kingfisher has staged something of a comeback, with the UK breeding population increasing year on year to 2014.

The hazards of the cold are not just the difficulties of finding food. One unfortunate Kingfisher settled on a metal pipe only to find that the frost 'glued' it to the metalwork. In its panic, the poor bird broke its leg in two places. It was rescued, but died later.

Floods also kill Kingfishers. Whole broods can be lost to breeding season floods and overflowing waterways may make it much harder to find a meal. Having said that, research by Čech and Čech published in 2013 found that breeding Kingfishers were able to adapt and cope with flood conditions, catching a greater variety of larger fish and feeding their young fewer, bigger fish so that the nestlings still achieved a good weight.

Food for others

It is said that the Kingfisher's bright plumage consists of warning colours that tell any would-be predator that they taste awful. If this is true, it does not always work.

The Sparrowhawk (*Accipiter nisus*) can grab prey in mid-air or take it from a perch, and they have certainly been known to take a Kingfisher. Ian Llewellyn watched a Kingfisher's response to a Sparrowhawk that had perched right above it. The Kingfisher went into a threat display, holding its vertical pose for about 15 minutes. Some predators are opportunists and one Kingfisher was in the wrong place at the wrong time when a Great Black-backed Gull (*Larus marinus*) flew over. The gull took the Kingfisher, flew off with it and perched, much to the disgust of John Gascoigne, the warden of a country park in southern England, and the birdwatchers that he was accompanying. The gull took to the air and gulped down its unusual meal.

Kingfishers are not totally safe from predators in the water either. 'My Halcyon River', a *Natural World* film

Below: Sparrowhawk and Great Black-backed Gull – both have been observed eating Kingfishers.

made by Charlie Hamilton James for the BBC, showed two angry female Kingfishers fighting in the water. As if from nowhere, an American Mink (*Neovison vison*) appeared and took one of the birds. It tried to take the second bird too, but failed. In Sweden Mink are known to predate Kingfishers at the nest and their feathers have been found in Mink spraint.

Ian Llewellyn also observed some Kingfisher behaviour that he believes was an adult checking the tunnel for predators: he saw a female proceeding down the gloomy tunnel swinging her wide-open bill from side to side.

Predators and Kingfishers

It is not only Mink that will raid a Kingfisher's nest. The Red Fox (*Vulpes vulpes*) does it, and so does the Brown Rat (*Rattus norvegicus*). The Brown Rat could be a significant predator of Kingfishers. Records show four nests, two in each of two successive years, on two rivers in Avon, England, all with full clutches of eggs, which were destroyed by rats. When one of the 'ratted' nests was investigated, the tunnel was clogged with soil, stones, moss, leaves and fish bones. Having eaten the eggs and the female Kingfisher the Rat had left little behind and had made her own nest in place of the birds'. All that remained of the female and the clutch was her bill, her left wing, some feathers, and some bits of eggshell. The Rat must have gained access to the tunnel by holding on to a small root. Predation by rats was also recorded by Rosemary Eastman. This time the rat dug its way into the nest chamber and took a female and her chicks. All that was left over after this attack was the lower mandible, a bit of skull and some feathers. Every part of the chicks seemed to have been eaten. It is possible that Weasels (*Mustela nivalis*) also take Kingfishers and Stoats (*Mustela erminea*) have also been listed as predators. One nesting attempt went badly wrong when a European Mole (*Talpa europaea*) burrowed into the chamber, presumably accidentally.

Above: Three Kingfisher predators: Mink, Red Fox and Brown Rat. The latter are known to have brought mass destruction to a nest chamber.

Above: Little remained of this female Kingfisher and her clutch of eggs after a Brown Rat had wreaked havoc in the nest chamber.

Domestic cats are often quoted as a significant predator of Kingfishers. Given the bird's lifestyle and the average cat's aversion to water it seems unlikely, but British Trust for Ornithology (BTO) ringing data suggest that it does play a part. Up to 1997, just over three quarters of the 737 ringed Kingfishers that had been found were dead. It was not always obvious what killed them, but a cause of death was given to 59 per cent of the birds. Of these, 16 per cent of deaths were attributed to 'domestic predators, mainly cats'. However, some of those records are presumably based on a cat showing up with a dead or dying bird. The cat may not have taken the bird – it may have picked it up after an unfortunate collision with a window or car.

Left: Cats do turn up with Kingfishers, but it can be hard to know whether they took the bird themselves, or found a dead or injured one.

Man-made hazards

Above: Some Kingfishers are killed when they are hit by cars.

Below: Sadly, and probably slowly, a lost hook and line killed this Kingfisher in Bristol in 2010.

Like some other birds, Kingfishers sometimes accidentally fly into windows, perhaps thinking that they have found a convenient shortcut. Others are hit by cars. BTO ringing data suggest that these are significant causes of death. Of the dead birds recovered where the cause of death was certain, 61 per cent of the fatalities were attributed to humans, directly or indirectly. Most of these deaths were in road traffic accidents or the result of flying into windows. A different kind of human cause led to the death of a Kingfisher in 2010. The dead bird was found inside a factory in Kent. It had been ringed just over three years earlier and its routine flight path took it through the factory. The building was also its roosting site. An ailing economy meant that the factory, which had previously been working all week long, had to be closed for part of the week. The Kingfisher was trapped in the building and died.

Most anglers fish responsibly, but discarded fishing tackle has been a threat to wildlife for many years. As the tragic picture (left) shows, victims include the Kingfisher.

Poison

The higher an animal is on the food chain, the more likely you are to suffer from a simple ecological principle called bioaccumulation, which sees toxins working their way up the chain and then into body tissues. Kingfishers are no exception, with chemicals such as pesticides from farmland finding their way into the waterways, and consequently into the Kingfisher via its food.

There are indirect consequences too. Pollution reduces fish populations, making Kingfisher food harder to come by. And, as climate change brings periods of drought, and our demand for water rises, less water remains in our waterways, making the toxins even more concentrated.

Above: Pesticides and other pollutants may cause problems for Kingfishers.

A survival strategy

Given the day-to-day hazards the Kingfisher must contend with, it is just as well they are able to produce so many young, with decent clutch sizes and the capacity for a second or even a third brood. One might think this would enable the species to make a relatively rapid comeback after a population setback, but this does not always happen. In the United Kingdom the BTO's Waterways Bird Survey, which ran from 1974 to 2007, showed a decline in Kingfisher numbers from 1974 that lasted about a decade. At that point the population started to increase again but numbers didn't return to their 1974 levels until about the year 2000. On a more positive note, there was a quick recovery after the catastrophe of the 1939–40 winter, when the Kingfisher population on a 109km (68 mile) stretch of the Thames had almost recovered after only three years, and 1972's *London Bird Report* tells of a speedy comeback after the cold winter of 1962–63 too.

Above: The impacts of climate change on our waterways and our own increased use of water may make things even tougher.

Past Imperfect; Future Tense?

Today the UK is home to an estimated 4,900 pairs of this glorious little bird. Its population goes up and down over time. It is currently on the Amber list of Birds of Conservation Concern and the BTO has not identified any long-term population trend. But without a bit of 19th-century bird protection legislation, things might have been very much worse.

Using Kingfishers as a weather forecasting tool (see page 114), which still took place as late as the early 1800s, was not good for the species, and neither was the 19th-century boyhood urge to kill one to add to the collection. Eastman talks of the possibility of the Kingfisher becoming extinct in England. That was written in 1969 and the author was wondering what might have happened without the 1880 Wild Birds Protection Act. The act

Opposite and below: A diving Kingfisher (opposite) and one about to take flight (below). Population sizes go up and down but, without Victorian bird protection legislation, Kingfisher numbers could have plummeted even further.

came into being because of the efforts of the Victorian naturalists and 'sentimentalists' who campaigned against the persecution suffered by numerous bird species. It gave the Kingfisher a degree of protection in the closed season (from 1 March until 1 August), though it remained 'fair game' during the open season. The 19th century had been a very tough time for the species.

Stuff it

Twenty years earlier, the bitter winter of 1859–60 had taken its toll on the Kingfisher population. Those that survived were not left to their own devices in the hope of a recovery. This was the Victorian era, and taxidermists were in demand. Stuffed birds were must-have accessories and that definitely included the Kingfisher, which became a victim of its own beauty. Taxidermists like to work with specimens that are in good shape, so to reduce 'damage' to their quarry, the bird catchers would work at nest sites and use a net or snare to take birds at the entrance. When they were taken the adults may well

Below: Stuffed birds don't look as good as the real thing.

have had eggs or dependent young in the nest. The market for Kingfishers was very buoyant – one taxidermist is said to have acquired 100 dead Kingfishers in just a year. For the stuffers, dead Kingfishers were the stuff of dreams.

Flaunt it

If you've got it, flaunt it, and Victorian ladies liked to flaunt bits of Kingfishers. Someone had to get those bits first of course – an 1864 account in *The Field* tells of someone shooting Kingfishers along the Thames for just this purpose. Once acquired, the specimens featured on hats and other items of 'finery', as feathers, a wing, or even the whole bird. Thankfully, tastes in fashion have changed.

Kill it

Its good looks made people want to shoot it, but it may have been the Kingfisher's supposed incompatibility with the fishing world that caused it the most problems. The Victorian game fishing world believed that the species' appetite for salmon and trout fry was just too big to accommodate and Kingfishers were ruthlessly persecuted. Stones were pushed into nest tunnels so that access was impossible. Kingfishers were shot. Traps were set including the cruel pole trap, which was outlawed in 1904. The pole trap is a spring-trap tethered to the top of a pole. When a bird lands on it, the jaws spring shut around the bird's legs and the bird dies hanging upside-down from the pole. Perhaps the ultimate cruel irony was that Kingfisher feathers were used to make fishing flies. Trade in Kingfishers continued into the 1920s – nets intended for wildfowl took a by-catch of Kingfishers, and each dead bird meant an extra shilling in someone's pocket.

Above: A fishing fly – in Kingfisher blue and Kingfisher orange.

Poison it

The industrial revolution changed the way we live. Unfortunately, it also poisoned our waterways, which would not have been good news for the Kingfisher. It has been suggested that the Kingfisher population in 19th-century Yorkshire may have suffered as a consequence.

Predict it

In *The Birds of Wiltshire* by the Rev. A. C. Smith, published in 1887, we are told that, up until the early part of the century, it was not unusual for Kingfishers to be hung from the ceiling as an aid to weather forecasting. The theory was that the dead bird would turn so that its bill pointed in the direction from which the wind was coming; in another story the bird becomes a compass and the bill points north.

Myth and legend

A French legend tells a story of how the Kingfisher acquired its wonderful colours. The legend says that on Noah's ark, the Kingfisher was a grey bird, and when Noah got fed up waiting for the dove to come back after it had been sent out to find somewhere dry he enlisted the services of the Kingfisher. But, to escape a storm, the Kingfisher flew very high and got struck by lightning, which gave the bird its blue upperparts. The bird then sought the heat of the sun, but misjudged the distance and singed his underparts, which turned orange, and his rump. When he got back, the flood had subsided and the ark was no more. Since then, the Kingfisher has been flying along rivers, searching and calling for Noah.

In Greek mythology, Alcyone or Halcyone drowned herself in the sea to be with Ceyx, her husband, who had died when his ship went down courtesy of one of Zeus's thunderbolts. Instantly, the gods transformed Halcyone and Ceyx into Kingfishers (some versions of the story say that Ceyx became a Gannet or tern). Their fragile nest floated on the sea, and for a week either side of the winter solstice, the sea was calmed while the eggs were brooded. These were 'halcyon days', and until about 200 years ago, sailor superstition said that the seas were calm when Kingfishers were raising their offspring. Today 11 Kingfisher species are in the *Halcyon* genus and 22 are in the *Ceyx* genus.

Above: After drowning herself Halcyone was transformed into a Kingfisher – according to Greek mythology, that is.

The *Halcyon* kingfishers are one genus of wood or tree kingfishers. Typically, they are large kingfishers that don't actually fish. *Ceyx* kingfishers are a genus of so-called 'river kingfishers'. They are small kingfishers that, contrary to their name suggests, are mostly found in forested areas. These aren't really fishers either.

Write about it

The Kingfisher has been a source of inspiration in a range of arts and media, from poetry to advertising. For Victorian poets the colours of the Kingfisher often influenced their work. In Edward Lear's *K was a Kingfisher* the bird is described as 'So bright and so pretty! – Green, purple and blue', while John Clare, in *The Kingfisher*, adopts a more comparative approach declaring that 'the peacock's tail is scarce as fine'.

Sell it

More recently, Kingfishers regularly feature in brand and business names. It is easy to understand why Kingfisher Environmental Services selected its name – the company specialise in water quality management and the Kingfisher is a bird that thrives in clean, unpolluted habitats. Is it the bird's superb diving skills or merely its association with water that led to the Kingfisher Leisure Centres that can be found dotted around the UK? The logic behind the numerous Kingfisher fish and chip shops is more obvious, though let's hope their menus include more than Minnow, Bullhead and Stickleback!

The rationale behind companies adopting Kingfishers within their branding is not always so clear. Indian restaurant aficionados will be familiar with Kingfisher beer, a brand of beer marketed by the United Breweries Group who also owned India's now defunct Kingfisher Airlines. One might speculate the latter name hinted at fast and direct flights but why the beer might have been named Kingfisher is less obvious.

In the UK the DIY company Kingfisher PLC, whose brands include B&Q and Screwfix, incorporate a stylised Kingfisher image into their logo. Kingfisher Building Products, a completely unrelated company, chose its name for three main reasons: the company's founder was Roger Fisher, who was full of admiration for the hard-working birds, and he recognised that the bird's striking blue-and-orange colours could become a powerful and memorable brand.

Above: Kingfisher beer – good with a curry!

On the up?

Right: Unfair competition.

It seems that towards the end of the 19th century things were improving for Kingfishers in Britain. At the northern edge of its UK range, there were nests in central Scotland and in 1903 there was one in Stuartfield, just over 32km (20 miles) north of Aberdeen. A few gentle winters and the 1880s act were believed to have contributed to the bird's change of fortune. With much less persecution, life for the 20th century Kingfisher improved, and populations grew in some areas. But there were setbacks, including the killer winters of 1946–47 and 1962–63.

In Scotland

BirdTrack is an online project of the BTO, the RSPB, the Scottish Ornithologists' Club and BirdWatch Ireland. It brings together sightings from lots of birdwatchers to find out more about where birds occur and their migration habits. BirdTrack records from 2008 showed the Kingfisher in Scotland at a similar latitude to the 1903 record mentioned above, but this masks what happened during the intervening years. Those cold winters sliced through the Scottish population and polluted rivers are thought to have hindered any comeback in much of the country. By the 1960s the Scottish range had contracted, with regular breeding reaching no further north than very southern parts of the country. Since then Kingfishers had been pushing north again, at least until the cold winters of 2009–10 and 2010–2011. It is likely that since then trends in Scotland have been similar to those for the UK as a whole, with the breeding population diminished by each cold winter, but then recovering progressively to 2014.

Elsewhere

Across the Irish Sea, the last quarter of the 19th century saw Kingfishers breeding throughout Ireland, but in many areas it was far from common. Information on 19th-century persecution of the species in Ireland is scant, but 1954's *Birds of Ireland* suggests that it was a serious problem for the bird. Things improved through much of the 20th century but then took a turn for the worse.

In mainland Europe, to some degree at least, the story is a familiar one, and many populations have been hit hard by pollution and persecution. But it is not all bad news. The Belgian population crashed to 190 pairs in the early 1970s but recovered and was estimated as 900–1,500 pairs in the early years of this century. There is good news from the south of Poland where numbers seem to be increasing – in the 1960s, Kingfishers in this area were virtually consigned to history. The Austrian population has also done well after a very sticky patch from 1950 to 1970.

Below: A healthy looking Kingfisher. The species' status in Scotland has improved since the 1960s, but the freezing winter of 2009–10 and 2010–11 could have hit the population hard.

The current state of play

Globally, the Kingfisher is doing OK and is currently regarded as of 'Least Concern' by BirdLife International. In Europe, however, things are not quite as positive. The 2015 European Red List of Birds categorises the Kingfisher as 'Vulnerable' because of a rapid drop in numbers across Europe – its population decline is estimated at 30–49 per cent over a period of about 13 years. The main threats have been identified as climate change and severe weather events, pollution and damage to its habitat.

Kingfishers can be found in most of England, much of Wales and some of Scotland, with a good scattering of records across Ireland too. The 2015 *Birds of Conservation Concern* puts the species on the Amber list. The good news is that it is not one of the birds that the conservation world is most worried about. It is, however, certainly one to keep an eye on.

To the future

The Kingfisher is a great bird to watch, but make sure that you do not disturb it as this is illegal. Loitering near a nest site can be particularly dangerous as it can deter the parents from going in to feed the young. Eventually, things are compounded because the young will fall silent if they do not get enough food or become chilled. When this happens the adults are not getting any 'feed me' messages, so they stop bringing food to the chicks. In the UK, the Kingfisher is a Schedule 1 species, which means that it is on the 'most protected list' of 1981's Wildlife and Countryside Act. Aside from the ethical issues, disturbing the blue blur is an offence, which could see the offender in jail for six months, facing a £5000 fine, or both.

If they are to thrive, Kingfishers need good quality habitat, with enough of the right fish and suitable nesting opportunities. Waterways need to be managed sensitively so that nest sites are not inadvertently destroyed by heavy machinery, and pollution needs to be kept under control

Kingfisher controversy

In 2015 something exciting and controversial happened in the conservation world. During an expedition to the Solomon Islands scientists caught a male Moustached Kingfisher (*Actenoides bougainvilli*). The man who found it, Christopher Filardi, had been looking for this species for over 20 years. The only records of it prior to 2015 were of a female that had been 'collected' in the 1920s, two more females from the 1950s, and one brief view. The bird that the expedition had caught was ornithology's first encounter with a male Moustached Kingfisher. After very careful consideration, the decision was taken to 'collect' the bird. This means that the bird was killed and became a scientific specimen.

It sounds shocking, but collecting has been the scientific norm for a long time. Some see specimens as the ultimate proof of a species' existence, and a very valuable resource for future research. Sacrificing this one bird, it is argued, might help the conservation world save the species. The scientists had concluded that there were enough Moustached Kingfishers in the area, and that the habitat was in sufficiently good shape, that taking one bird would have no detrimental effect on the population as a whole.

Not everyone agreed and, unsurprisingly, there was outrage.

so that the birds can enjoy some decent water. We need to use water wisely – drainage can damage Kingfisher habitat.

Climate change means we are likely to experience more extreme weather and this could include icy winter spells, floods and droughts, all of which can make life difficult for the Kingfisher. It is also possible forest fires will have an adverse effect on some Kingfisher populations. So always do your bit to minimise the impacts of climate change.

Below: Common Kingfishers live in many parts of the world and, at a global level, are not a major cause for conservation concern. Long may it stay that way.

How to See a Kingfisher

This chapter gives you some practical advice that, hopefully, will help you to find and see Kingfishers. We have also included an appendix that lists some places around the UK that you can go to to try to see this wonderful bird.

Brief encounters

Many people's experience of the Kingfisher is a snatched glimpse of turquoise flying away and out of sight. They can fly fairly fast – estimates suggest speeds of at least 40–45 kilometres per hour (25–28 miles per hour) on a still day. To spot them more efficiently, try to learn their call, a piercing and high-pitched whistle. If you hear it, scan low over the water for a bird in flight.

Opposite: No colours, but unmistakably still a Kingfisher.

Below: A wonderful shot of Kingfishers in fading light.

How to find them

Above: Hides can enable close encounters with Kingfishers.

Below: Scan possible perches for a view of the Kingfisher. Their blue plumage may betray their presence, but they are not always as conspicuous as this bird.

Find out where Kingfishers occur in your area – you may not have to travel very far. Check the local bird club's website – look for information about birdwatching sites or lists of recent bird sightings – either of these could point you in the right direction. Look in county bird reports to see where Kingfishers have been seen, or use a 'Where to Watch Birds' book to find a site with Kingfishers. Or just ask a local birdwatcher. If that fails, get a map, look for some likely looking patches of water and go and explore.

Think about timing. Kingfishers can be seen throughout the day and at any time of the year, though some sites only have them in the winter. Your chances are better at some times than others, however. I have seen Kingfishers very well early in the morning and they are especially obvious during February and March as the breeding season gets underway, and in the autumn when youngsters are being driven out of the parental territory. There are more around in the autumn too because of all the young birds that are out and about and these can be much more approachable than the adults.

Learn their calls and use your ears. That is how you will discover most of the Kingfishers you see. You can listen to them for free online – try the RSPB's website

Left: A clue for the Kingfisher detective. Patches of bright, white droppings can help you locate a bird's favourite perch.

(rspb.org.uk/kingfishers), for example. Their calls will help you locate the bird – often the sighting will be of a blue blur flying low over the water. Alternatively, there are a number of apps available which include recordings of bird songs and calls.

Check likely looking perches by the waterside, including distant ones. Scan the trees at the edge of a lake, for example. The wider, shallower parts of a river may harbour more fish – check these carefully.

Look for Kingfisher poo patches on banks or trees. Fresh droppings are an oily yellow colour but when they dry they become very white and conspicuous. The patches can be as big as a dustbin lid. With practice you might be able to find pellet spots too. Then sit and wait – unless it is the breeding season. If the perch is near the nest your presence could cause problems, so keep your distance. Remember that the Kingfisher is a Schedule 1 protected species.

If you have found a safe place to sit and watch, try using a portable hide or make the most of natural cover. If the bird sees a bit of movement or picks out a human shape that could be the end of your Kingfisher encounter.

Follow the Birdwatchers' Code. You can find it on the RSPB's website (rspb.org.uk/birdwatcherscode).

Enjoy your time with this brilliant bird!

Glossary

Cloaca The external opening of a bird's gut, urinary system and reproductive system.

Covert Small feathers that cover the bases of the main wing feathers and tail feathers (upper wing-coverts, upper tail-coverts etc) and the ear openings (ear-coverts).

Crustaceans A group of invertebrates that includes crabs, lobsters, crayfish and woodlice.

Fledgling A fledged bird that has grown its first set of 'proper' feathers.

Genus The first part of a plant or animal's scientific name. Birds that are in the same genus are closely related. The Common Kingfisher is in the *Alcedo* genus.

Home range The area a bird normally occupies. Unlike a territory, home ranges are not defended.

Juvenile A bird is a juvenile from the time it fledges until its first moult.

Mandible The upper and lower parts of a bird's beak.

Molluscs A group of invertebrates that includes snails, slugs and octopuses.

Monogamy/monogamous Simplistically, having only one mate at one point in time – being part of a pair. However, some 'monogamous' birds also mate with birds other than their main partner.

Taxonomic order The animal kingdom is divided into phyla, the phyla into classes, the classes into orders, the orders into families, the families into genera (singular = genus), and genera are divided into species.

Pectoral fins The pair of fins attached to the sides of a fish at the front.

Polygamy Having more than one mate at any one point in time. A male with more than one female is polygynous. A female with more than one male is polyandrous.

Preening Using the bill to help keep the plumage in good condition.

Primary feathers The large flight feathers attached to the outermost part of the wing.

Red List A list of threatened species.

Ringing Putting a ring on a bird's leg. Individually numbered rings enable bird movements and lifespans to be studied if a ringed bird is caught again or found dead. Coloured rings enable individual birds to be identified without recatching them.

Syndactyl Where two toes are joined along part of their length.

Territory An area that a bird defends.

Tyndall effect The scattering of light by particles in its path.

Acknowledgements

Thank you to all those who have played a part in making this book what it is. Firstly, those who studied Kingfishers and made the results accessible, particularly David Boag, Ron and Rosemary Eastman and Ian Llewellyn. Thank you Ian for all of the time that you gave me. For information and 'consultancy' thanks are due to Norman Crowson, Mark Boyd, Karen Guest, Anne Cotton at BTO Scotland, and Gordon Grainger at Kingfisher Building Products.

At Bloomsbury, thank you Julie Bailey and Katie Read for all that you did to bring this book into being. And at the RSPB, thanks to Ben Andrew and Kate Smith for your helpful contributions.

Without its photos, this book would be much less attractive and informative – thank you to all the photographers who provided the images, not least Ian Llewellyn. Finally, a big thank you to Anna MacDiarmid, who project-managed the book through its layout stages, Liz Drewitt for the proofreading, and Susan McIntyre – whose design skills have been key to making *Kingfishers* look as good as it does.

Resources

KINGFISHER PLACES

This list is a useful starting point if you would like to see Kingfishers in the UK.

England

Bedfordshire

Priory Country Park, near Bedford.

Cambridgeshire

Milton Country Park, Cambridge.
Paxton Pits, near St Neots.

Dorset

Radipole Lake and/or Lodmoor RSPB Nature Reserves, Weymouth.

East Yorkshire

Blacktoft Sands RSPB Nature Reserve, near Goole.

Essex

Fingringhoe Wick, 5 miles from Colchester (Essex Wildlife Trust).
Lee Valley Country Park, 2 miles north of Waltham Abbey.

Gloucestershire

Slimbridge (Wildfowl & Wetlands Trust).

Greater London

Brent Reservoir, north London.
Rainham Marshes RSPB Nature Reserve, near Purfleet.

Greater Manchester

Pennington Flash Country Park.

Hertfordshire

Rye Meads, near Hoddesdon (Hertfordshire and Middlesex Wildlife Trust/RSPB).
Tring Reservoirs.

Lincolnshire

Whisby Nature Park, near Lincoln.

Norfolk

Cley, 4 miles north of Holt (Norfolk Wildlife Trust).
Strumpshaw Fen RSPB Nature Reserve, east of Norwich.
Titchwell RSPB Nature Reserve, 5 miles east of Hunstanton.

Northamptonshire

Summer Leys Local Nature Reserve, near Wellingborough (Bedfordshire, Cambridgeshire, Northamptonshire and Peterborough Wildlife Trust).

North Yorkshire

Wheldrake Ings, 8 miles from York (Yorkshire Wildlife Trust).

Nottinghamshire

Colwick Country Park, near Nottingham.

Oxfordshire

Farmoor Reservoirs, near Oxford.

South Yorkshire

Old Moor RSPB Nature Reserve, near Barnsley.
Potteric Carr, near Doncaster (Yorkshire Wildlife Trust).

Suffolk

Lackford Lakes, north-west of Bury St Edmunds (Suffolk Wildlife Trust).
Minsmere RSPB Nature Reserve, near Eastbridge.

Warwickshire

Brandon Marsh, near Coventry (Warwickshire Wildlife Trust).
Kingsbury Water Park, Sutton Coldfield.

West Sussex

Chichester Gravel Pits.

West Yorkshire

Fairburn Ings RSPB Nature Reserve, near Castleford.

Wiltshire

Cotswold Water Park.

Worcestershire

Upton Warren, near Bromsgrove (Worcestershire Wildlife Trust).

Scotland

Angus & Dundee

Balgavies Lock, Forfar (Scottish Wildlife Trust)

Edinburgh & the Lothians

Basinch and Duddingston, Edinburgh (Scottish Wildlife Trust)

Glasgow and Clyde Valley

Falls of Clyde, Lanark (Scottish Wildlife Trust)
Lochwinnoch RSPB Nature Reserve, Paisley
Barons Haugh RSPB Nature Reserve, Motherwell

Wales

Conwy

Conwy RSPB Nature Reserve.

Carmarthenshire

Llanelli Wetland Centre (Wildfowl & Wetlands Trust).

Pembrokeshire

Welsh Wildlife Centre, south of Cardigan (Wildlife Trust of South and West Wales).

BOOKS

Boag, D., *The Kingfisher*, Blandford, 1982 and 1988.
Boag, D., *The Kingfisher*, Shire Natural History, 1986.
Eastman, R., *The Kingfisher*, Collins, 1969.

WEBSITES

BTO: www.bto.org
RSPB: www.rspb.org.uk

Image credits

Bloomsbury Publishing would like to thank the following for providing photographs and permission to reproduce copyright material.

While every effort has been made to trace and acknowledge all copyright holders, we would like to apologise for any errors or omissions and invite readers to inform us so that corrections can be made in any future editions of the book.

Key t = top; l = left; r = right; tl = top left; tcl = top centre left; tc = top centre; tcr = top centre right; tr = top right; cl = centre left; c = centre; cr = centre right; b = bottom; bl = bottom left; bcl = bottom centre left; bc = bottom centre; bcr = bottom centre right; br = bottom right

AL = Alamy; FL = FLPA; G = Getty Images; NPL = Nature Picture Library; RS = RSPB Images; SH = Shutterstock

Front cover Andrew Howe/G, t, Lisa Geoghegan/G, b; **Back cover** Mark Hughes/G, t, Loop Images/G, b; **spine** Lisa Geoghegan/G; **1** Mark Hughes/G; **3** Ian Llewellyn; **4** Ian Llewellyn; **5** Klaus Echle/NPL; **6** Ernst Dirksen/Buiten-beeld/Minden Pictures/FL; **7** Ian Llewellyn t, b; **8** Ian Llewellyn; **9** Andrew Mason/FL, t, Ohmori/G, b; **10** Ian Llewellyn, t, b; **11** Ian Llewellyn; **12** Ian Llewellyn; **13** Thomas Hinsche/G, t, Ian Llewellyn, b; **14** Chris Fredriksson/AL; **15** hfuchs/SH, t, DEA / BELLANI/G, b; **16** Andrew Astbury/SH; **17** Simon Bennett/Minden Pictures/FL, bl, aaltair/SH, br; **18** Robin Bush/G; **19** Martin Willis/FL, t, Martin B Withers/FL, b; **20** Steve Young/FL, t, Dave Montreuil/SH, b; **21** Neil Bowman/FL, tl, b, Ignacio Yufera/FL, tr; **22** Richard Du Toit/Minden Pictures/FL, t, RAJU SONI, b; **23** Alan Murphy/Minden Pictures/FL, t, Auscape/G, cr, jdm.foto/SH, b; **24** BGS_Image/SH, t, Thomas Cockrem/AL, b; **25** Jurgen & Christine Sohns/FL, t, Carlton Ward/G, b; **26** Andrew M. Allport/SH, t, Ian Llewellyn, b; **27** Marek CECH/SH; **28** Dave Montreuil/SH; **29** Tony Heald/NPL; **30** Karel Bartik/SH; **31** Kajornyot Wildlife Photography/SH, t, Bird Hunter/SH, b; **32** Ian Llewellyn; **33** Ian Llewellyn; **34** Ian Llewellyn, t, David Hughes/SH, c, The National Trust Photolibrary/AL, b; **35** FotoRequest/SH; **36** Ian Llewellyn; **37** Erni/SH; **38** FotoRequest/SH; **39** Paul Hobson/FL; **40** Mike Powles/FL; **41** Laurent Geslin/G, t,

Jerome Murray – CC/AL, b; **42** Ian Llewellyn; **43** Ian Llewellyn; **44** Ian Llewellyn; **45** stormcab/SH; **46** Martin Pelanek/SH, tl, boyphare/SH, tr, Michal Masik/SH, bl, AlekseyKarpenko/SH, br; **47** Ian Llewellyn, t, c, b; **48** Ian Llewellyn; **49** Paul Sawer/FL; **50** Charlie Hamilton James/NPL; **51** Andrew Astbury/SH, t, Ian Llwellyn, b; **52** Ian Llewellyn; **53** Ian Llewellyn, tl, tr, bl, br; **54** CHC3537/SH, bl, Jausa/SH, br; **55** image BROKER/AL, t, Paul van Hoof, Buiten-beeld/Minden Pictures/FL, b; **56** Ian Llewellyn; **57** Paul Sawer/FL; **58** RogerPowell/NPL, t, ThomasHinsche, BIA/Minden Pictures/FL, b; **59** Ian Llewellyn, t,b; **60** Ian Llewellyn; **61** Heinz Hudelist/G; **62** Erni/SH; **63** htu/G, t, Ian Llewellyn, b; **64** Ian Llewellyn; **65** Carl Morrow Photography; **66** Ian Llewelyn; **67** Bernd Zoller/G; **68** Dean Bricknell/RS; **69** Frederic Desmette/Biosphoto/FL; **70** Neil Burton/SH; **71** Ian Llewellyn; **72** blickwinkel/AL; **73** Ian Llewellyn; **75** Ian Llewellyn, t, aaltair/SH b; **76** Frank Hecker/AL; **77** Ian Llewellyn; **78** Ian Llewellyn; **79** Ian Llewellyn, t, b; **80** David Kjaer/G; **81** Bruno Guerreiro/EyeEm/G; **82** Ian Llewellyn, bl, br; **83** Ian Llewellyn, bl, br; **84** Ian Llewellyn; **85** Scott M Ward/SH; **86** Matej Ziak/SH; **87** Angelo Gandolfi/NPL; **88** David Boag/AL, t, Paul Tymon/SH, b; **89** Ian Llewellyn, t, Albert Visage/FL, b; **90** Ian Llewellyn; **91** Michel Poinsignon/NPL, t, b; **92** Charlie Hamilton James/NPL; **93** David Boag/AL t, Angelo Gandolfi/NPL b; **94** Ian Llewellyn; **95** Angelo Gandolfi/NPL; **96** Angelo Gandolfi/NPL; **97** Ian Llewellyn, t, Erni/SH, b; **98** dreamnikon/SH, t, b; **99** Paul Sawer/FL; **100** Charles Hamilton James/NPL; **101** Ian Llewellyn; **102** cowboy54/SH; **104** Ian Llewellyn, t, Wayne Hutchinson/FL, b; **105** Art Wittingen/SH, t, Bildagentur Zoonar GmbH/SH, b; **106** Mike Lane/FL, t, Richard Guijt Photography/SH, c, Bildagentur Zoonar GmbH/SH, b; **107** Ian Llewellyn, t, blickwinkel/AL, b; **108** Ben Schonewille/SH, t, Ian Llewellyn, b; **109** Nick Spurling/FL, t, ShaunWilkinson/SH, b; **110** Ian Llewellyn; **111** kajornyot wildlife photography/SH; **112** Jeremy Pembrey/AL; **113** Johann Helgason/AL; **114** public domain/ Wiki commons; **115** Bloomberg/G; **116** Erni/SH; **117** Ian Llewellyn; **119** Ian Llewellyn; **120** Ian Llewellyn; **121** Ian Llewellyn; **122** Fred van Wijk/AL, t, Ian Lewellyn, b; **123** Ian Llewellyn

Index

INDEX